AGRICULTURAL HERITAGE

NEW INDIA PUBLISHING AGENCY
New Delhi-110 034

About the Authors

Prof. (Dr) A Zaman is an internationally renowned scientist specialization with agricultural water management having a spectacular career of 40 years. His lifetime and original research contribution in the field of irrigation water management, rainfed agriculture, cropping system, watershed management, dry land agriculture and biodiversity conservation, to name a few, globally appreciated leading to outstanding teachers in farm universities in India and abroad. A large number of original concepts, strategies, techniques a tool those developed which have contributed enormously to agricultural science and benefitted the global society.

Prof Zaman is working as the Director, School of Agricultural Sciences, Sister Nivedita University as on and from 1st May 2020. Prior to his joining to the post he was Founder In-charge, Agriculture (School of Agriculture and Allied Sciences), The Neotia University on and from 1st June, 2018. He was Emeritus Professor in Agriculture in M.S. Swaminathan School of Agriculture, Centurion University of Technology and Management (September 2016 to 31 May 2018) and Programme Coordinator, World Bank funded West Bengal Accelerated Development of Minor Irrigation Project implemented by Water Resource Investigation and Development Department (WRIDD), Government of West Bengal (September 2014 to August 2016).

Dr. Zaman worked as Head, Irrigation Expatriate Team, Government of Uganda for a 2-year period (1994-1996) under ITEC Programme being deputed by the Government of India to design, layout and installation of famers managed small scale irrigation for sustainable crop cultivation and more than 150 peer-review research papers and 179 publications in the proceedings of national and international seminars and conferences held in the country and abroad in his credit and having authorship of 10 text books in agriculture published with several book-chapters. Dr Zaman guided and supervised 7 (seven) Ph D and more than 50 (fifty) students towards postgraduate thesis work.

Dr. Md. Hedayetullah PhD is an Assistant Professor/ Scientist, AICRP on Chickpea, Directorate of Research, Bidhan Chandra Krishi Viswavidyalaya, Kalyani, Nadia, West Bengal. Acting as an Officer-In-Charge under All India Coordinated Research Project on Chickpea. He was former Agronomist, National Bank for Agriculture and Rural Development, Balasore, Odisha, India. Former Assistant Professor, M.S. Swaminathan Institute of Agriculture Science, Centurion University of Technology and Management, Gajapati, Odisha, India. Dr. Hedayetullah was an Assistant Professor, College of Agriculture, Tripura, Government of Tripura, India. Dr. Hedayetullah acquired a Bachelor of Science (Agriculture) degree from H.N.B. Garhwal University, UK, India. He received Master of Science (Agronomy) from Palli Siksha Bhavana, Institute of Agriculture, Visva Bharati University, Sriniketan, West Bengal, India. He received PhD (Agronomy) from BCKV, Mohanpur, Nadia, West Bengal, India. He was awarded the Maulana Azad National Fellowship Award from the University Grant Commission, New Delhi, India. He has received several fellowship grants from various funding agencies to carry out his research works during his academic career. Dr. Hedayetullah is author and co-author of 30 research papers, 5 review papers 26 book chapters and 7 books.

He guided 2 students for postgraduate degrees and supervising 1 research students for postgraduate degrees in Agronomy. He has attended more than 30 National and International Seminar, Symposium, Conference, Workshop and training courses. He is Principal Investigator and Co-PI of many ICAR and Ad-hoc projects. He is an Editorial Board Member of Amity Journal of Agribusiness (AJAB). He is the life members of the CWSS, B.C.K.V, W.B, India; and Indian Society of Pulses Research and Development and Fire Ask Sobuj. He is a Member of International Institute Bengal Basin, Barkley, USA.

AGRICULTURAL HERITAGE

A. Zaman
Director, School of Agricultural Sciences
Sister Nivedita University
Kolkata-700 156, West Bengal

Formerly
Consultant to World Bank, Minor Irrigation Project
Head, Irrigation Expatriate Team, Government of Uganda, East Africa
Head, Department of Agronomy
Dean, Faculty of Agriculture,
Chief Scientist, Water Management,
Director of Research
Bidhan Chandra Krishi Viswavidhayalaya
Nadia-741 235, Mohanpur
West Bengal, India

Md. Hedayetullah
Assistant Professor
Department of Agronomy
I/c AICRP on Chickpea
Bidhan Chandra Krishi Viswavidhayalaya
Nadia-741 235, Mohanpur
West Bengal, India

NEW INDIA PUBLISHING AGENCY
New Delhi-110 034

NEW INDIA PUBLISHING AGENCY
101, Vikas Surya Plaza, CU Block, LSC Market
Pitam Pura, New Delhi – 110 034, India
Email: info@nipabooks.com
Web: www.nipabooks.com

For customer assistance, please contact
Phone: + 91-11-27 34 17 17 Fax: + 91-11- 27 34 16 16
E-Mail: feedbacks@nipabooks.com

ISBN: 978-93-95763-93-6

Composed and Designed by NIPA.

Preface

India is facing deep trouble with fall of expected economic growth and production mainly due to sharp decline in revenue generation and collection under world-wide pandemic from COVID-19 (Novel Corona Virus). The agriculture sector is the only ray of hope which still maintaining the positive gross domestic product (GDP). An adequate understanding of the challenges that exist in the new millennium consist of the challenge to establish good management practices in a geographically dispersed environment, to establish social security policies and procedures that adequately reflect the organizational context and new business processes, to establish correct structures of responsibility, given the complex structuring of organizations and information processing activities and to establish appropriate information technology disaster recovery plans keeping parity with past and present. In this context, to study the agricultural heritage of the country has got paramount importance.

The book entitled "Agricultural Heritage" is written as a textbook for the students in agriculture in the all agricultural universities, agricultural colleges at the undergraduate level as per syllabuses of 5th Dean Committee's recommendation taking account of lesson-wise 18 lecture plans derived from the content of the prescribed syllabus. In the past decade, the agricultural productivity increased substantially due to intensive management and introduction of fertilizer and irrigation responsive high yielding varieties of cereal crops to the fore. But it is essential to know that was where the agriculture started from the ancient past and that it reached to modern agriculture now as of today over a thousand of years in the process of civilization, social changes, and evolution.

Though a few books are available on the title those did not have an arrangement of class lectures in accordance with the prescribed syllabus to give a readymade and easy guidelines to the teachers for imparting lesson to the students. The course is being offered within the 1-3rd semester in undergraduate level within the curriculum. Hence, the attempt has been taken in arranging the prescribed syllabus systemically. This might be considered as the remedial measures removing the technical difficulties in bringing out a complete series of lesson plans to meet the demand.

The authors are indebted to the sources from where the information collected and made fervent appeals to the scientists and researchers to convey their suggestions on improvement of the documents making it valid and more valuable.

September 20, 2020 **Authors**

Contents

Preface .... *vii*

1. Introduction to Agricultural Heritage, Definition of Heritage, Agricultural Heritage, Scope of Agriculture and Need to Study Agriculture Heritage 1

Introduction 1
Definition 2
Agriculture heritage 3
Scope of agriculture heritage 5
Objectives of the agriculture heritage study 8
Need to study agricultural heritage 9
Remarkable characteristics 12
Conclusion 14
Some pertinent short questions 15

2. Relevance of Heritage to Present Day Agriculture 16

Introduction 16
Agriculture and colonialism 18
Indian agriculture after independence 18
Krishi Panchang 19
Pinching making 20
Characteristics of *Krishi Panchang* 20
Resilience of Indian agriculture 20
Major crops and yields 22
Conclusion 25
Some pertinent short questions 25

3. Past Status of Agriculture and Farmers in Society Specific Role of/in Ensuring Food Security in Farming Systems Since Ancient Periods 26

Introduction 26
Agriculture in early historic period 26
Maurya period and onwards 27
Land and soil 27
Crops cultivated 28
Cereals 28
Fruits and Spices 28

Legumes 28
Green Vegetables 29
Dye Crops 29
Plant protection 29
Pests 29
Diseases 29
Others 29
Horticulture interventions 30
Animal husbandry 31
Concealed farms were of two types: large and small 32
Fishery 32
Poultry 33
Sericulture 33
Farm management 33
Meteorology and agriculture 33
Farm area and its management 35
Farm plan 35
Achievement system 36
Indus civilization period 36
Crop cultivation 38
Horticulture 38
Wild animal and Livestock 39
Fishery 39
Farm tools and implements 39
Meteorological studies 40
Conclusion 41
Some pertinent short questions 41

4. Present Status of Agricultural Advancement, Modern Agricultural Education and Role of Agri-Societies 42

Introduction 42
Definition of an agricultural society 43
Social order 43
The groups in agricultural society 44
Agricultural cooperative and marketing society 45
Concept of producer organization 46
The capital appearance 46
Modern agricultural education 47
Emerging challenges in agriculture sector 48
Career scope of agricultural science and technology 49
Career opportunities in agriculture science & technology 50
Advancements in agriculture during five year planning periods (1951-2011) 51
Conclusion 57
Some pertinent short questions 58

5. Agricultural Resources Available, Importance of Agriculture and Scope of Agriculture 59
Introduction 59
Irrigation Projects 62
Development of status 64
Idea of origin 65
Premise of ownership 66
Individual resources 66
Community owned resources 66
National resources 66
International resources 66
Land utilization 67
Land use pattern 68
Soil as a resource 69
Renewable and non-renewable resources 69
Conclusion 72
Some pertinent short questions 72

6. Importance of Farmers and Ancient Agricultural Practices and Scientific Basis, Status of Agriculture as Advice by Sages to Kings on Their Duties Towards Farmers 73
Introduction 73
Farmer in village administration 74
Kings, Sages and Saints 77
Agricultural heritage and society 83
Conclusion 85
Some pertinent short questions 86

7. Heritage of Crop and Water Management in Ancient and Pre-historic and Medieval Period 87
Introduction 87
Irrigation system 87
History of irrigation 89
Use of irrigation water 92
Objectives of irrigation application 94
Rivers and water resources management 95
River associated streamlets serving the purpose of irrigation 97
Irrigation projects 98
Efficient water use 98
Irrigated agriculture 99
Efficient water management technologies 99
Use of poor and marginal quality water for irrigation 101
Water management 102
Conclusion 102
Some pertinent short questions 103

8. Journey of Indian Agriculture and Its Development from Past to Modern Era 104
Introduction 104
Indices to analyze age-old agriculture 106
Chronological agricultural technology development in India 107
Probable age-old crop introductions 114
Crops of Indian origin 123
Conclusion 126
Some pertinent short questions 127

9. Crop Production Through Indigenous Traditional Knowledge (ITKs) 128
Introduction 128
Documentation of ITKs 129
Possible stakeholders of the ITKs 129
Categories of ITKs 129
ITKs for crop cultivation 130
Preservation of ecological knowledge 131
Needs for documentation 131
Role of conventional knowledge 132
Indigenous knowledge 134
Crop production constraints 135
Use of indigenous knowledge in farming 136
Conclusion 141
Some pertinent short questions 142

10. Plant Disorders, It Causes, Symptoms, Treatment Materials and Plant Protection in Ancient India Through Indigenous Traditional Knowledge (ITKs) 143
Introduction 143
Spatio-temporal variations 144
Present arts of knowledge 145
ITKs in plant disease control 147
ITKs in Insect-pest superiority 148
Popular ITKs against crop pests 148
Mechanical control practices 150
Formulation of control measures 152
Control of Stored Grain Pests 155
Conclusion 156
Some pertinent short questions 156

11. Crop Voyage In India and The World 157
Introduction 157
The hypotheses of the crop voyage 157
History of civilization and agricultural technology 159

Influences on advancement of Indian agriculture 160
Liaison and relationship between East and West 161
Contrarians...161
Crop voyage and domestication 162
Major crops grown in India 162
Factors affecting crop diversification 164
Consequences of changes in crop pattern... 165
Success story in crop diversification 166
Production and productivity effects.. 167
Conclusion 170
Some pertinent short questions...170

12. Medicinal Plants and Thier Importance in Agricultural Heritage.. 171

Introduction 171
Definition of medicinal plants 172
Plants for human health.. 172
Use of plants as medicines 173
Herbal medicines...174
Sources of medicinal plants 174
Constraints and opportunities...175
Spreads of medicinal plants in Eastern Region 175
Social perspective 176
Environmental perspective 177
Potential of equitable commercialization 177
National trade and enterprise development 178
World market trends..178
Domestication and cultivation 178
Agro-technology 180
Opportunities in developing the medicinal plant's sector 180
The way forward 180
Institutional support..181
Stakeholder's participation...181
Marketing perspective.. 182
Existing policies 182
Link with bio-partnership and rural communities with the industry 184
Conclusion.. 186
Some pertinent short questions... 187

13 Gender Equity in Ancient and Modern Agriculture 188

Introduction 188
Women in agricultural heritage 191
Women authorization 192
Self-facilitated teams of female 192
Innovation of SHGs.. 19

Women's contribution to agricultural production 197
Conclusion 198
Some pertinent short questions 199

14. Crop Significance and Crop Classifications 200
Introduction 200
Causes of crop diversification 202
Opportunities in advanced agro-technologies 203
Strategies for crop diversification 203
Crop classifications 205
Ancient World 216
Miscellaneous 220
New World 220
Conclusion 221
Some pertinent short questions 222

15. National Agricultural Setup in India 223
Introduction 233
National Agricultural System 223
1. ICAR System 223
2. Agricultural University System 236
3. Other agencies imparting agricultural education 241
Conclusion 244
Some pertinent short questions 245
References 245

16. Current Scenario and Journey of Indian Agriculture and Its Impacts 246
Recent Agricultural Production Scenario 246
Capital Formation in Agriculture and Allied Sectors 251
National Policy for Farmers (NPF), 2007 251
Doubling the farmer's income 252
Rashtriya Krishi Vikas Yojana (RKVY) 252
Objectives of RKVY Project 253
Sub-Schemes launched under RKVY 254
RKVY performances 255
National Food Security Mission (NFSM) 256
Outcome of NFSM 257
National Mission on Oilseeds and Oil Palm (NMOOP) 257
National Mission for Sustainable Agriculture (NMSA) 260
Rainfed Area Development (RAD) 261
Soil Health Management (SHM) 261
Pradhan Mantri Krishi Sinchayee Yojana (PMKSY) 262
Paramparagat Krishi Vikas Yojana (PKVY) 263
Agricultural Technology Management Agency (ATMA) 264

Kisan Call Centres (KCCs) 265
Strengthening of IT Apparatus in Agriculture and Cooperation in the States and Union Territories (AGRISNET) 268
National Crop Insurance Programme (NCIP) 268
National Horticulture Mission (NHM) 269
Horticulture Mission for North East and Himalayan States (HMNEH).. 270
National Agro-forestry and Bamboo Mission (NABM) 270
Coconut Development Board (CDB) 271
National Horticulture Board (NHB) 271
Conclusion 273
Some pertinent short questions 274

17. Indian Agricultural Concerns and Future Prospects Agricultural Vision, The Future and Challenges Ahead 274
Introduction 274
Speculation on Indian agriculture 274
Rainbow revolution 275
Per drop more crop and income 275
Opening up of the markets 276
Future Research and Development 277
Gap existing that required that specific intervention(s) 278
The major components of the project taken up 279
Implications of Projected Outlook 284
Conclusion 285
Some pertinent short questions 286

18. International Agricultural Research Setup 287
Introduction 287
Major avenues of international collaboration 287
Global Agricultural Research System 288
CGIAR (Consultative Group on International Agricultural Research) ... 288
International Agricultural Research System 291
CGIAR Gene Banks 289
Other International Centers 300

References 307

1

Introduction to Agricultural Heritage Definition of Heritage, Agricultural Heritage, Scope of Agriculture and Need to Study Agriculture Heritage

Introduction

The specific agricultural systems and landscapes have been created, shaped and maintained from generation to generations of farmers and herders based on diverse natural resources, using locally adapted management practices. Indian heritage is unique and prestigious than any other civilization in this world. We must feel proud being a citizen of India for its rich cultural heritage. Agriculture in India has a long history from Neolithic age of 7500-4000 B.C. People changed their life style from nomadic hunter of wild berries to cultivator of land. Agriculture is gifted from the wisdom and teachings of great saints. The knowledge achieved and techniques adopted have been passed down through generations. The traditional farmers have developed the eco-friendly farming systems and mixed farming, mixed cropping, crop rotation etc type of systems were practiced by them. The older generations of the farmers of India passed the depth of knowledge and experience of great epics of ancient India.

Indian agriculture heritage is known as unique and prestigious than any other civilization in this world. Its rich cultural heritage in India has a long history from Neolithic age of 7500-4000 B.C. People changed their life style from nomadic hunter of wild berries to cultivator of land. Agriculture is gifted from the wisdom and teachings of great saints. The knowledge achieved and techniques adopted in generation to generations. The traditional farmers have developed the eco-friendly farming systems and mixed farming, mixed cropping, crop rotation etc type of systems were practiced but the older generations of the farmers of India passed the depth of knowledge and experience of great epics of ancient India and its consequence has come up in present day agriculture.

The preservation of wild animals was maintained and hunting as a sport by the ruler was considered as detrimental to proper development of the character and personality according to *Manusmriti*, 2nd Century BCE. In the *Puranas* (300-750 AD), the names of the experts in animal husbandry were found like *Shalihotra* on horses and *Palakapya* on elephants. More than 75 plant species was mentioned in the four Vedas. *Satapatha Bhrahamna* mentioned over 25 species, and more than 320 plants were found in *Charkaa Samhita* (300 BC) an Aayurvedic treatise. Over 750 medicinal plant species were recorded in Susruta (400 BC). Rigveda (4000 BC), the oldest book, referred a large number of poisonous and non poisonous aquatic and terrestrial, and domestic and wild creatures and animals. About 500 species of plants were written in Puranas. In the Rigveda, it has been mentioned about the hundreds and thousands of cows, horse drawn chariots and race courses where chariot races were held; camels yoked to the chariots; sheep and goats offered as victims for worship, and the use of cloths made from wool. The famous Cow *Sukta* indicated that the cow also became the basis of rural economy. *Sukta* is also worshiped as the mother of the *Vasus*, the *Rudras* and the *Adityas*, and also the pivot of Immortality. *Rigveda,* which is the most ancient literary work of India, believed that Gods were the foremost amongst agriculturists. According to the thesaurus, "*Amarakosha*" which was written by the Jain or Buddhist scholar *Amarasimha* in Sanskrit, Aryans were agriculturists. Agriculture, cattle rearing and commerce were the essential subjects of the king that must be learnt. This was prescribed by Manu and Kautilya. The most ancient literature available in India was "Vedas" which includes the four Vedas (rig, yajur, sama, atharvana), nineteen Brahmanas.

Definition

Agricultural heritage is refers to the values and traditional practices adopted during ancient agriculture, which are relevant to the present day agriculture system. In a word, agricultural heritage denotes the values and traditional practices adopted in ancient India, which are more relevant for present day system where Indian agricultural heritage is unique than any other civilization.

Globally important agriculture heritage system is defined as "Remarkable land use systems and landscapes which are rich in globally significant biological diversity evolving from the co-adaptation of a community with its environment and its needs and the aspirations for sustainable development" (FAO, 2002).

In the 1600's the first Dutch settlers of Boerum Hill found themselves amidst Native American maize fields. Christina Kelly came across Reginald Bolton's book, "Indian Paths in the Great Metropolis". There she found detailed maps of Brooklyn's historic maize fields. After tracking down heritage seeds consistent with those of the period, she replanted that field with the cooperation of Jeff Hutcheson at what is now the corner of Smith and Bergen in Brooklyn's Boerum Hill neighborhood.

Agricultural diversity is embedded in traditional diets of indigenous peoples around the world. Preserving indigenous knowledge is in effect, saving the world's agricultural diversity that is quickly disappearing from the overwhelming rise of modern monocultures.

Agriculture heritage

Traditional systems of agriculture represent an accumulative heritage of world initiated since the Neolithic of elementary importance. Stylish agriculture constantly threatens the property of this inheritance. Because of their ecological and cultural significance and conjointly the wealth and breadth of accumulated knowledge and knowledge inside the management and use of resources that these systems represent, it's imperative that they be thought-about globally very important resources to be protected and preserved, still as allowed to evolve. Policy support and actions at international, national and indigenous levels are needed to allow GIAHS to evolve whereas providing continuing product and services in their totality and integrity.

Agricultural heritage denotes the ethics and acceptable practices adopted in age old India, which are added accordant for present day system. List of available documents on agriculture during age-old and medieval period as follows:

1. *Rigveda* (c.3700 BC)
2. *Atharvaveda* (c. 2000 BC)
3. *Ramayana* (c.2000 BC)
4. *Mahabharata* (c.1400 BC)
5. *Krishi-Parashara* (c.400 BC)
6. *Kautilya's Artha-sastra* (c.300 BC)
7. *Amarsimha's Amarkosha* (c.200 BC)
8. *Patanjali's Mahabhasya* (c.200 BC)
9. *Sangam abstract* (Tamils) (200 BC-100 AD)
10. *Agnipurana* (c.400)
11. *Varahamihir's Brhat Samhita* (c. 500 AD)
12. *Kashyapiyakrishisukti* (c.800Ad)
13. *Surapala's Vrikshayurveda* (c.1000 AD)
14. *Lokopakaram by Chavundaraya* (1025 AD)
15. *Someshwardeva's Manasollasa* (1131 AD)

16. *Saranghara's Upavanavioda* (c.1300 AD)
17. *Bhavaprakasha-Nighantu* (c.1500 AD)
18. *Chakrapani Mistra's Viswavallbha* (c.1580 AD)
19. *Dara Shikoh's Nuskha Dar Fanni-Falahat* (c.1650 Ad)
20. *Jati Jaichand's dairy* (1658-1714 AD)
21. *Anonymous Rajasthani Manuscript* (1877 AD)
22. *Watt's Dictionary of Economic Articles of India* (1889-1893 AD)

Formation of Indian sub continent Pangaea, the super continent 250 million years ago the Earth's seven continents were all aggregate calm into a cool abstemious alleged 'Pangaea'. This huge cool abstemious was amidst by one gigantic ocean alleged *Panthalassa. Laurasia* about 180 years ago the cool self-disciplined *Pangaea* began to breach up in the Mesozoic Era into *Laurasia* and Gondwanaland. *Laurasia* was fabricated of the present day continents of Arctic America known as Greenland, Europe, Angara acreage (Asia) in Russia, Siberia and China in the north. Gondwanaland was fabricated of the present day continents of South America, Africa, India, Australia, and Antarctica. At this time India was not affiliated to Asia. The huge ocean of *Panthalasa* remained. 'The Triple Junction' was formed because of a three-way breach in the band acceptance massive foundation flows in three cautions and caked out bedrock over hundreds of aboveboard far of Africa and South America. The rocks in these two continents were produced at the aforementioned time and in the aforementioned place. This tells us that South America and Africa were affiliated together at one time. **South America** is believed to have been joined with **Africa** from the late Paleozoic Era to the early Mesozoic Era, until the supercontinent **Pangaea** began to rift and break apart about 225 million years ago. Today these two continents are far by the Atlantic Ocean. *Laurasia* was still moving, and as it confused it bankrupt up into the continents of freezing America, Europe and Asia (Eurasian plate). In the additional stage, the *Gondwanan* continents far from anniversary added during the Jurassic and Cretaceous period in the backward Jurassic, South America far from Africa. This created addition basin amid these two continents. The eastern bank of Arctic America far from the Moroccan appendage of Africa, the breakdown of the *Gondwana* land opened up the Atlantic and the Indian Ocean. The Indian Subcontinent confused hundreds of far in 135 million years at a plentiful acceleration about 4 inches per year. The Indian bowl exhausted into the Eurasian bowl (Asia) with such acceleration and force created the tallest great quantity on Earth, the Himalayas. The Tethys Ocean was accepting uncomfortable out of actuality in the east of the Alpines as India approached Asia. Concrete cartography of Indian sub-continent finds a lot of outstanding actuality about the concrete cartography of India that is accustomed analysis of the country with three audible segments of absolutely antithetical character: (i) the Himalayas, the plentiful arrangement to the north, (ii) the Indo-Gangetic alluvial plain of freezing India extending from the Punjab to Assam, and (iii) the Peninsula of the Deccan to the south of the Vindhyas, a solid enduring block of the earth's crust, copiously composed of some of the a lot of age-old rocks, which the acknowledgment of ages that has carved into a amount of large quantity in ranges, plateaus, valleys and plains.

Inherent to the concept of **globally important agricultural heritage (GIAHS)** is acknowledgement that indigenous knowledge has inherent profit, and holds development potentials. Accidentally in many components of the developing world, there still exists a diversity of indigenous and ancient practices of theme management, beside systems of selection management, and soil and conservation. Many rural peoples according to World Health Organization are resource-poor farmers, are inventively self- dependent, and continuously carried on with the experiment, adaption and initiation. The agricultural communities living in ancient agricultural landscapes in globally important agricultural heritage sites might hold many of the potential answers to the challenges of agricultural production and natural resources management in an era of worldwide global climate change. The GIAHS framework acknowledges that there are real opportunities for building on theme and living diversity and finance in indigenous communities and their resources, indigenous knowledge and institutions, to loosen hunger and impoverishment in rural areas, rather than relying on excessive external inputs and sometimes inappropriate and unsustainable technologies from outside.

Scope of agriculture heritage

The ancient farmers developed the natural farming systems in many diversity i.e. mixed farming, mixed cropping, intercropping and crop rotation. The ancient India was possessed with rich natural resources and conveys the deep knowledge to the present day agriculture. Agriculture heritage as authenticated by the Food and Agriculture Organization (FAO) are remarkable land use systems and landscapes development evolving from the co-adaptation of an association with its atmosphere and its needs and aspirations for acceptable development. The specific agricultural systems and landscapes accepted worldwide. The landscape created, shaped and maintained by ancestors of farmers and herders based on assorted accustomed resources and application locally climate management practices. Building on surrounded knowledge and experiences, these able agricultural systems reflect the change of humankind, the development of knowledge, and its technological accord with nature. These systems accepted the result in outstanding landscapes and adjustment of globally convincing agricultural biodiversity, indigenous knowledge systems. Millennia communities of farmers, herders, fishers and wilds humans accepted developed complex, diverse, and locally acclimatized agricultural systems. These systems have been accepted and managed with time-tested, able combinations of techniques and practices that usually led to association with security, and the absorption of familiar assets and biodiversity. Agricultural heritage systems can still be start on throughout the ancient periods about 5 million hectares, which accommodate a basic aggregate of social, cultural, ecological and economical casework to humankind. Agricultural biodiversity, well-ventilated ecosystems and an accepted cultural heritage all these systems sustainably accommodate various accessories and services and alimentation guidance for millions of poor and small and marginal farmers. The actuality of plentiful GIAHS testifies to the skillfulness and expertise of humans in their use and management of the natural resources, biodiversity and ecosystem dynamics, and able use of concrete attributes of the landscape, codification in acceptable but evolving knowledge, practices and technologies. Whether accustomed or not by the truthful community, these affiliated agricultural systems aggregate the foundation for alongside each other and approaching agricultural innovations and technologies. Their cultural, ecological and agricultural development is still self-evident in various locations of the world, maintained as different systems of agriculture. Through an arresting action of co-evolution of Humankind and Nature, globally important agricultural heritage accepted as emerged over centuries of cultural and biological interactions and synergies, which accumulate adventures of rural peoples. Indian agriculture began by 9000 BCE as an aftereffect of indigenous agriculture of plants, and conduct of crops and animals. Settled movement presently followed with accessories and techniques accepting developed for agriculture. Double monsoons during the ancient period led to two harvests accepting reaped in one year. Indian articles immediately accomplished the formed absolute trading networks and adopted several crops.

Globally important agricultural heritage shows the historic cultures and knowledge. These cultures, settlements, landscapes and habitats have suffered dramatically inside the wake of business and agricultural revolutions and advances of science, technology, commerce and communications inside the 19th and 20th century's. The few that conjointly survive as flag bearers of the earlier traditions are worth safeguarding as a vicinity of the protection of the world cultural and natural heritage. Agricultural heritage landscapes are not exclusively very important landmarks of historical worth but together rely upon living and evolving agricultural communities. These communities are the custodians of institutional, ecological and cultural heritage that has a diffusion of benefits and services at the indigenous, national and international levels.

Inherited traditions of farmers draw numerous natural resources and pattern domestically. These systems region unit managed with tried and true, ingenious mixtures of techniques and practices that have generally gave idea to community food security, and conjointly the conservation of natural resources and selection. These globally important agricultural heritages have resulted not exclusively in outstanding landscapes of having aesthetic beauty but maintenance of worldwide very important agricultural selection, resilient ecosystems and a valuable cultural heritage. Above all, these systems sustainably offer multiple products and services, food and living security for several poor and marginal farmers. The existence of numerous globally important agricultural heritages around the world testifies to the power and ingenuity of people in their use and management of finite resources, variety, theme dynamics, and ingenious use of physical attributes of the landscape, written in ancient but evolving knowledge, practices and technologies. Whether or not it is recognized by the scientific community, these ancestral agricultural systems represent the inspiration for contemporary and future agricultural innovations and technologies. The cultural, ecological and agricultural diversity remains evident in many components of the world, maintained as distinctive systems of agriculture. Agricultural heritage favors importance for the provision of indigenous food security, high levels of agro-biodiversity and biological diversity, store of indigenous technological knowledge. The biophysical, economic and socio-cultural resources have evolved beneath specific ecological and socio-cultural constraints to make outstanding landscapes. The samples of such agricultural heritage systems are inside the loads of and are home to thousands of ethnic groups, indigenous communities and indigenous populations with a numberless of cultures, languages and social organizations.

Plants and animal considered capital to their edition by the Indian that came to be admired and acclaimed average ages proverb on irrigation knowledge a new similar of self-possession in India and Indian crops affecting the economy of added regions of the under Islamic patronage. Land and water management systems were developed with an aim of compatible growth. Despite some stagnation during the after forward-thinking era, India was able to advance a complete agricultural program. Agriculture heritage is different than any other civilization in any part of the world. Indian experiences an appreciative affluent cultural heritage. Agriculture in India is not of hatred origin, but has a continued history dating back to Neolithic age of 7500-4000 B.C. It afflicted the activities of appearance in indigenous man from traveling hunter of agrarian people and roots to agriculturalist of the specific land. Agriculture is benefited from the sharpness and commentary of plentiful saints. The good judgment acquired and practices adopted accept been anesthetized down through generations. The acceptable farmers accept developed the attributes pleasant agriculture systems and practices such as alloyed farming, alloyed cropping, crop rotation. Plentiful documents of age-old India sponsor the deep hole of knowledge that bedeviled by the earlier ancestors of the farmers. Need and accent for belief agricultural heritage that has lot of affiliated acceptable practices anesthetized from one bearing to added generation and as well agriculture. In India, agriculture is not an occupation but it is a way of movement for affluent Indian populations. Hence the present day bearing should be acquainted about our age-old and acceptable agricultural systems and practices. This will accredit to body the approaching analysis action also. India has fabricated amazing advance in agriculture and its allied fields, but the tone of voice on accelerated use of inputs after because of their adverse continued term base has created several problems accompanying to sustain knowledge of agriculture. Irrational use of organic fertilizers, insecticides and analysis of familiar possessions is aggressive the agro eco-systems. Soil is accepting indigent, water and air accepting natural resources attenuated and there is an accumulation scratch of these resources including biogenetic resources. Therefore, incorporation of these resources is now living to acceptable composition of agriculture. The indigenous technological knowledge (ITK) provides old age practices for controlling insect, pest and diseases into the acceptable agriculture, because these innovations accept been agitated on from one bearing to addition as an ancestors technology. Indigenous technological knowledge also facilitates the experience in new comer in agriculture which is very much key factor for taking emergency decision in unpredicted agriculture situation. There are several examples of admired acceptable technologies but horribly these small and marginal systems are dying out. It is sensitive that to collect, certificate and examine these technologies so that the precise principle/basis could be appropriately understood. Once this done, it will be easier to add, clarify and advancement them by aggregate with the advanced correct technology.

Objectives of the agriculture heritage study

1. To know agriculture that is way of movement, advancement and progress not only an activity;
2. To access acquaintance of the prosperous heritage of Indian agriculture which is different than any other civilization;
3. To implant a faculty of pride amidst the people, decidedly agricultural acceptance to agriculture that has been acceptable practices for generations;
4. To activate truthful analysis based on adequate technology;

Need to study agricultural heritage

Indian agriculture has lot of inherited sustainable practices those are orally passed from one generation to other generation. Agriculture is not only an occupation but it is the backbone of the society and the way of life for majority people of India. Hence it is important to the present day generation to become aware about our ancient and traditional agricultural systems and practices. This will help us to develop the future scope of research strategy also.

India has made tremendous improvement in agriculture and its allied sectors, but due to the intensive use of inputs without considering their adverse effects to the environment on a long term basis which has created several problems related to sustainability of agriculture. Non-judicious application of chemical fertilizers, insecticides and irrational use of natural resources is threatening the agro eco systems. Soil is degraded, water and air are polluted and plant and animal genetic resources are in danger. Therefore, importance is now shifted to sustainable form of agriculture.

The indigenous technical knowledge (ITK) is the valuable information for the sustainable agriculture, because these technologies have been carried on from one generation to another as a family technology. There are several examples of valuable traditional technologies in India but unfortunately these small local systems are going to extinct day by day. It is imperative that the agricultural scientist should collect, document and analyze these technologies so that the scientific principle/basis behind them could be properly understood.

Agriculture heritage is need to study as because basic guidelines for agricultural growth and development in specific agro-eco system of the country which is very much importance in present day agriculture. Some points on old-age systems are highlighted and discussed below:

1. Indian millet terrace agro-ecosystems

Major as well as minor millets are suitable in terrace systems with integrated forest use. These systems together embrace numerous agricultural choices and various elements: as an example, integrated rice-based systems (e.g. rice-fish culture, rice-fish-duck, rice-fish-taro) with numerous rice and fish varieties/ genotypes; and integrated forest, land and water use systems, notably found in East Asia and conjointly the chain.

2. Multiple cropping/multistoried cropping systems

These are exceptional mixtures and/or plantings of numerous crop varieties with or whereas not integration of agro-forestry. They are characterized by resourceful microclimate instruction, soil and water management schemes, and accommodative use of crops to touch upon climate knowledge. In agricultural heritage, maize and root crop based agro-ecosystems developed by the Aztecs in Mexico.

3. Understory farming systems

These are agricultural systems pattern combined or integrated life science, grove or various crop systems with every over story-canopy and understory environments. Farmers use understory crops to produce earlier returns, diversify crops/products and/or build economical use of land and labor. These practices are common inside the tropics, e.g. in taro-based or root cropping systems, planted along with various endemic plant varieties from indigenous genetic resources. These are common in island New Guinea, Vanuatu, King Islands and various Pacific small island developing countries.

4. Mobile and semi-nomadic pastoral systems

These are the rangeland/pastoral systems supported accommodative use of pasture, rangeland, water, salt and forest resources, through quality and variations in herd composition in harsh non-equilibrium environments with high animal genetic diversity and outstanding cultural landscapes. These embrace highland, tropical and sub-tropical dry land and arctic systems like Yak-based pastoral management in Ladakh and conjointly the high Tibetan upland in state and China; very intensive ground use in components of Mongolia and Yemen; cattle and mixed animal primarily based mobile pastoral systems, like of the Maasai in East Africa; reindeer-based management of Champaign of the Saami and Samoyed inside the temperate forest areas of Scandinavia and nation. The landscapes intentional by these systems typically offer habitats for wild species alongside species.

5. Ancient irrigation, soil and water management systems

These are the ingenious and finely tuned irrigation, soil and water management systems commonest in dry lands, with a high diversity of crops and animals practiced such as: (i) Use of ancient underground water distribution systems evolution comes in the state and various central Asian countries with associated home gardens and endemic shade fish species living in underground waterways; (ii) The oasis of the Maghreb inside the deserts of North Africa and conjointly the Sahara; (iii) Ancient valley foundation and soil management just like the water management systems in lake, the Niger basin and interior delta e.g. floating and flooded rice systems; and (iv) Various inventive irrigation systems in Bamileke region, Cameroon; of Dogon tribes in French Sudan and Diola tribes in Senegal; still as a result of the village tank system in Sri Lanka and state.

6. Advanced multi-layered home gardens

These agricultural systems feature advanced multi-layered home gardens with wild and domesticated trees, shrubs and plants for multiple foods, medicines, ornamentals and various materials, in all knowledge with integrated agro-forestry, Sweden fields, looking gathering or placental, just like the house garden systems in China, India, the Caribbean, and the Amazon.

7. Under water level

These agricultural systems feature soil and water management techniques for creating productive land through debilitating delta swamps. The systems perform throughout a context of rising ocean and watercourse levels whereas endlessly raising land levels, thereby providing a multifunctional use of land i.e. for agriculture, recreation and business, nature conservation, culture conservation and urbanization e.g. Polder or dyke systems inside the Netherlands; Kuttanad wetlands in Kerala, India; floating gardens in state and South Asia.

8. Grouping agricultural heritage systems

These systems feature various grouping agricultural practices and techniques of managing soil, water and crop cultivars in sloping lands from higher to lower valleys pattern mixed and/or a combination of cropping systems and group action indigenous knowledge systems e.g. Seethampheta in Andhra Pradesh, the *Apatani* rice fish culture in Arunachal Pradesh, the Zabo system of rainwater harvesting, and jhum cultivation in Darjeeling inside the chain, and much of different systems in various states.

9. High-value crop and spice systems

The high value crop production was in list of ancient fields and high significance crops and spices. The production techniques adopted in such a ways to keep the crop in rotation or in intercropping system e.g. Saffron systems in Malaysia followed by several state and geographic regions in India.

10. Hunting-gathering systems

These systems feature distinctive agricultural practices like gathering of unbroken rice in Chad (Central Africa) and honey gathering by forest domicile peoples in Cameroon Central and other nation. There are numerous various agricultural heritage systems around the world meriting identification, assessment and dynamic conservation. One in each of the foremost tasks of the GIAHS partnership initiative is that this adds collaboration with indigenous communities, national governments and various national and international institutions.

Remarkable characteristics

By fostering associate in nursing current, dynamic conservation of leaders agricultural systems and sites that show distinctive agricultural landscapes around the world, a technique will emerge that gives tangible international services, whereas providing very important support to rural communities through enhancing food security, conservation and property use of selection, and maintenance of cultural identity. The distinctive ancient farming systems present at the GIAHS sites symbolize that at an equivalent time exhibit exceptional choices of worldwide and indigenous significance.

1. High levels of selection that play key roles to speed up theme functioning and together in providing theme services of indigenous and international significance. GIAHS systems typically replicate moneyed and globally distinctive agricultural selection displayed at the sphere and together at the landscape level forming the premise for food production systems. A salient feature of GIAHS is their high degree of plant diversity inside the sort of rotations, poly-cultures and/or agro-forestry patterns. This strategy of minimizing risk by planting several species and styles of crops stabilizes yields over the long run, promotes diet diversity and maximizes returns even with low levels of technology and restricted resources. Genetic diversity provides security to farmers against diseases, pests, droughts and various stresses. It together improves knowledge of the cropping systems, permits farmers to use altogether totally different soil varieties and microclimates and derive multiple biological process edges and various uses from genetic variation among the species. At the landscape scale, diversification happens by group action multiple production systems.

2. Agro-ecosystems nurtured by cognitive comfortable systems and farmers' innovations and technologies wherein indigenous peoples living in GIAHS sites typically possess a broad content of the particulars of indigenous and complex ecology. This data regarding plants, animals, soils and jointly the overall setting has accumulated through a long series of observations transmitted from generation to generation. Indigenous farmers are aware that biological diversity might be an important think about generating ecological services, and inside the conservation of the resource base and foods on it they believe. Women, specially, are holders of much more pleased so play an essential role inside the conservation and utilization of selection in the process of agricultural heritage.

3. Resourceful systems and technologies of selection, land and water resource management and conservation which will be accustomed improve management of up to date agro-ecosystems. By searching for

ancient systems, scientists can learn plenty of knowledge regarding the dynamics of advanced systems, notably regarding the links between agricultural selection and theme perform and thereby contribute to the enrichment of the ecological theory and derive principles for usage inside the fashion of up to date property farming systems. as an example, in deciphering but intercropping follow works, farmers can create the foremost of the knowledge of cropping systems to use their own hold on nutrients. This data are gleaned to spice up during which throughout which farmers can manage soil fertility. Similarly, there could be plenty of progress in gadfly management schemes if the biological mechanisms at intervals the advanced structures of ancient agro-ecosystems are determined, so minimize crop losses because of insect pests, diseases and weeds.

4. Distributed agricultural systems that contribute to indigenous and national food and living security. Most small farming systems are productive, economical and property compared to larger farms despite their low use of chemical inputs. As a result of the exclusively resource-base offered for small farmers is on natural resources and their human capital. So that they diversify their genetic resources, they diversify their production systems and their sources of monetary gain, and each one this builds resilience. This contributes to food production, but together to environmental health, to the property of the natural resource-base so to the property of livelihoods. Small farms that manufacture grains, fruits, vegetables, fodder, and animal material inside an equivalent field are plenty of productive than big farms if the whole output is taken into consideration rather than yield from one crop. The yield blessings of distributed farming systems can vary from 20 % to 60 % on top of monocultures. Poly-cultures generally cut back losses because of weeds, insects, and diseases and build plenty of economical use of the offered resources of water, light, and nutrients. Ancient multiple cropping systems offer the most quantity as 20 % to 40% of the world's food provide for the people.

5. Farming systems that exhibit resiliency and robustness to handle disturbance and alter (human and climatic-environmental) minimizing risk inside the inside of knowledge. Many GIAHS farmers cope and even steel oneself against world global climate change, minimizing failure through increased use of drought-tolerant indigenous varieties, water harvesting and collection, intensive planting, mixed cropping, agro forestry, wild plant collection and a series of different ancient farming system techniques. Observations of agricultural performance on extreme climatic events inside the last 20 years have disclosed that

resiliency to climate disasters is closely connected to levels of farm selection. Many indigenous management practices that safeguard agro-ecosystems from climate variation embrace incorporation of constant and indigenous varieties into the agricultural system and increasing the temporal and concept of crop diversity of every at the sphere and landscape level. These points out the need to re-evaluate indigenous technology as a key provider of information on knowledge centered on the selective, experimental and resilient capabilities of ancient farmers in managing world global climate change and various external changes.

6. Systems that offer indigenous, regional and international theme services. The maintenance of high selection levels at GIAHS sites contributes to agricultural productivity and property through the theme services that selection provides. Agro-ecosystem perform is optimized via complementary interactions that emerge from added species in various agro-ecosystem, *i.e.* by mixture specific genotypes of crops for malady resistance, alongside as an example a legume species that may increase number seven inputs and exercise or by intercropping to support plenty of insect enemies with specific roles in leading pests. In many agriculture heritage sites agro-forestry systems are multifunctional operative landscape, giving kind of theme services and environmental edges like carbon sequestration, selection conservation and soil enrichment. In many regions, the management of various agriculture at intervals landscapes provides essential watershed functions, like maintaining water quality, management water flow, recharging underground aquifers, mitigating flood risks, sediment flows, and sustaining recent species and ecosystems.
7. The knowledge of ecological systems to produce product and services critically suppose rural communities having and sustaining numerous and complex styles of system, behavior, settlement, cluster membership and identity, gender relations, leadership and political organization, culture, modes of production, labour allocation, and technologies and practices. These replicate adaptation to and management of advanced social-ecological systems.

Conclusion

Knowledge in the subject on agricultural heritage is required to sustain and capitalize globally important information as it's a necessity to spice up understanding of the threats that they face, and confirm ways that during which to mitigate risks of land degradation, and conjointly the perverse impacts of process and international modification. Throughout to this sense, to prevent a lot of degradation of the system, their dynamic nature ought to first be recognized.

Their resilience depends on the potential to adapt to new challenges whereas not losing their biological and cultural wealth, and productive knowledge. Making an attempt to conserve globally important agricultural heritage system by "freezing them in time" would sure lead to their degradation and condemn their communities to impoverishment. The initiative emphasizes that globally important agricultural heritage is not regarding the past but it's regarding the future", referring to the approach centred on of us, human management and knowledge systems. This encompasses their socio-organization, economic and cultural choices that underpin the conservation and adaptation processes of agricultural heritage, providing support whereas not compromising their resilience, property and integrity. Building on indigenous knowledge and knowledge, these ingenious agricultural systems replicate the evolution of world, and its profound relationship with nature. These systems have resulted not exclusively in outstanding landscapes, maintenance and adaptation of worldwide very important agricultural selection, indigenous knowledge systems and resilient ecosystems, but, above all, inside the sustained provision of multiple products and services, food and living security for several indigenous individuals' members and indigenous peoples, well on the way aspect their borders.

Some pertinent short questions

1. Define agricultural heritage in the context of agriculture.
2. What are the objectives of study of agricultural heritage?
3. Mention five old age documents of agricultural heritage in our country.
4. What is full form of GIAHS?
5. Define grouping in agricultural heritage systems.

2

Relevance of Heritage to Present Day Agriculture

Introduction

In India, the history of agriculture dates embarrassed to Indus natural depression Civilization Era and even before that in some locations of Southern India. Today, India ranks more familiarized in surface area output. Agriculture and related sectors like biology and fisheries accounted for 137% of the value concerning 500th of the manpower in 2013. The livelihood buildup of agriculture to India's value is steady breakdown with the country's broad-based livelihood growth. Still, agriculture is demographically the broadest and highest livelihood across and plays a strong role within the all-inclusive socio-economic attachment of India. India exported $38 billion service of scientific discipline on-line symbols, authentic it the seventh larger scientific discipline business person used to and therefore the sixth larger web businessperson, plenty of its agriculture exports serve developing and developed nations as in 2013. Indian agricultural/horticultural foods squarely determined and exported to further more than 120 countries, primarily within the geographical area, environmental region, SAARC countries, the EU and United States. As per FAO (2010), agriculture statistics India is that the world's larger agent of in great quantity of fruits and vegetables, milk, spices, base crops that as jute, staples as millets and vegetable oil seed. India is that the further larger agent of wheat and rice in the world.

India stratified within the world's five larger producers of over 80th of scientific order ending things, as well as in large numbers banker's invoice crops feels affection for unusual crops including cotton, in 2010. India is one in every of the world's five larger producers of farm animal including dairy products and plenty of meat, with one in every of the quickest advanced rates, as of 2011.

It was claimed India's group is growing quicker than its facility to after-effects rice and wheat further aversion studies confirmation that will calmly supplement its growing population, further after-effects wheat and rice for all-round exports, if it will alleviate basal spoilage, advanced its basement and accretion its surface area wealthiest to those ready by further developing countries like Brazil and China. In year of June 2011, with an acclimatized weaken season, Indian agriculture ready best ceremony attainment of 85.9 million metric tons of wheat, 6.4% getting from a year earlier. Rice production in India hit a brand new ritual at 95.3 million metric ton, a seven times more than that of earlier year. Lentils and other crops grow in large area further staples attainment as well the over year. Indian farmers concerning 71 kg of wheat and 80 kg rice for each person availability in 2011. The per capita availability of rice each year in is currently conservatory than the per capita availability of rice each year in Japan. India exported $39 billion ritual of scientific disciplines on-line writing of documents in 2013, authentic it the 7th larger scientific discipline businessperson worldwide, and therefore the 6th larger web businessperson. This represents very small growth was concerning $5 billion as in 2004 web address. India is that the quickest growing businessperson of scientific discipline on-line writing documents over a 10-year period amounting its $39 billion of web exports which is further many folded the accumulated exports of the European Union (EU-28). India become one in every of the world's larger suppliers of rice, cotton and wheat.

India exported concerning 2.0 million metric tons of wheat and 2.1 million metric tons of rice in 2011 to Africa, Nepal, People's Republic of Bangladesh and further regions concerning the planet. Cultivation and attached work is amidst the quickest growing industries in India. Amidst 1990 and 2010, the Indian turn seizure fall doubled, whereas cultivation fall tripled. In 2008, India was the world's 6th larger agent of dreadful and new seizure of fisheries and therefore the further larger cultivation farmed curve as producer, India lags way back in acceding of super schedule incorporation at 200th that is to be tackled by authentic gettable super fragment wash out on-line symbols like eggs, meat, fish and chicken at cheaper costs.

India has credible a constant acceptance in the kilograms produced per hectare for some agricultural items, over the stand for 60 years. These assets acquire arise mainly from India's promising revolution, ambulatory passageway and proficiency address in infrastructure, ability of assets generation and reforms. Despite these alongside each other happenings, agriculture has the abeyant for affluence and complete achievement gains, because crop yields are still just 30% to 60% of the best adequate crop yields attainable in the farms of developed and other developing countries. Because of an all-embracing calendar chaos, agriculture brings in a calendar year through eFarming across a person could get more vegetables and fruits complete from Khet/Farm. Hence, agricultural heritage got paramount importance in the country as well all over having and local significance.

Agriculture and colonialism

Over 2500 years ago, Indian farmers had credible and voluminous agriculture prosperous species of sugarcane used to grow. It was in India, amidst the sixth and four BC, that the Persians, followed by the Greeks, convincing the highly praised "reeds that after-effects proceeding of honey from bees". On their reception journey, the Macedonian soldiers agitated the "honey address reeds," appropriately over extension other crops along with sugarcane in agriculture. India invented, by about 500 BC, the activity to after-effects sweat crops. Before the 18th century, agriculture of sugarcane was in large quantities cultivated in India. A few merchants began to adjustment in sugar, prosperity and a big-ticket cream exported in Europe until the 18th century. Sugarcane became broadly adapted in Europe in 18th century, and afresh accelerating to become a terrible alarm in the 19th century all over the world. This change of after-effect and address for sugar crops are found as a basal accumulation unleashed in livelihood and performed adequate changes. Sugarcane does not be plentiful in cold, frost-prone climate; therefore, adjoining and semitropical colonies were not found suitable to grow sugarcane. Sugarcane plantations became authoritarian of affluent and afflicted terrible migrations in 19th century and ancient 20th century as evident from documents available from Africa and from India, both influencing the ancient mix, political conflicts and cultural change of Caribbean, South American, and Indian Ocean and Pacific Island nations.

Indian agriculture after independence

In the years before independence, India has forged enormous higher requirement come up with food security. Indian population has tripled, and food-grain realization additional than quadrupled. There has been an affluent acceptance in achievable food-grain per capita. Before mid-1960s India relied on imports and received help to accommodate tranquil requirements of food grains. However, within two years of in 1965 and 1966, India alleviates its agricultural activity and increased food grain substantially and could not ahead on adopted support and imports for food security. India adopted strong activity reforms focused on the hunger of food grain and turn in self-sufficiency. This ushered in India's agricultural revolution changed agricultural scenario in the country. It began with the adaptation to acquire high yielding, dwarf wheat varieties in accumulated with bigger agriculture expertise to advanced agricultural productivity. The emphasis of Punjab led India's promising commitments and sufficient the precision of accepting enhancement of the country's food basket. The forward acceptance in succession was centered on the irrigated areas of the states of Punjab, Haryana and western Uttar Pradesh. With the farmers and the government admiral assimilation on land coverage prosperity and expertise transfer, India's complete food basket attainment was satisfactorily fulfilled.

Indian wheat land produced a boilerplate of 0.8 tons/ ha in 1948, produced 4.7 tons/ha of wheat in 1975 from the same land. Such accelerated advanced in land privileged circumstances enabled India to become food supply chain friendly by the 1970s. It as well empowered the smallholder farmers to seek supplementary assistance from the government department to acceptance of new agricultural technology for staple food produced per hectare.

Krishi Panchang

Krishi-Parashara Panchanga/ Krishi Panchang, an almanac (calendar) for the benefit of farmers in India, is prepared strictly in conformity with the ancient agricultural Sanskrit text *Krishi-Parashara*. This text is said to have been written around 4th century BC. The almanac is prepared for farmers to plan their agricultural operations during the year 2002-2003 (*Chitrabhani Samvatsara*). There may be some differences in some regions of India and the corrections may have to be made accordingly by the farmers. The views of readers on such regional variations are most welcome so that these views and suggestions can be included in the next year's almanac. But all farmers in India can generally use this almanac with local adjustments if any needed. This almanac can be used as a ready reckoned by farmers who do not have any knowledge of *Jyothishya* (Astrology), but who follow the Christian calendar. It is prepared for the first time and the entire almanac is based on the models (sutras) as available in the text.

The researcher developed the *Krishi panchang* (or) Agro ceremony (or) Agro-panchang. It may be authentic as basal astro-agricultural adviser calendar arise annually, giving annular caution on different aspects of agricultural and allied activities, basically suggesting amphitheatre wise, assessment practiced crop wise. Crop activity based on astro-meteorological prediction, giving advantageous time for chance different land accompanying operations, alternating with a ceremony for bold religious rites, festivals, celebratory fasts and some non-astrological agricultural guidance, primarily advantageous for the agriculture communities and bodies accepting incorporation in agricultural development. The adaptation of the proposed *Krishi Panchang* can broadly be categorized into two kanor groups as follows:

1. Information which changes every year;
2. Ceremony date and Holiday calendar;
3. Month – adroit circadian adviser for the able year;
4. "Rashiphal", i.e., month-wise forecasting of bodies accepting acclimatized zodiac sings;
5. Daily/monthly/annual acclimatizes forecasting for the authentic year;

6. Crop diplomacy of that year based on all-embracing positions;
7. Season-wise crop activity based on advancing weather; and
8. Information which charcoal irrespective of any valid year.
   (i) Theories applicable to agricultural and meteorological forecasting;
   (ii) Auspicious moments for agricultural and allied activities; and
   (iii) Some familiar agricultural guidance.

Pinching making

Simple **technique** of making pots by crudely shaping a ball of clay and then, by forcing the thumb into the centre, gradually **pinching** out the walls to an even thickness and the desired shape.

Characteristics of *Krishi Panchang*

The adequate and advantage of the proposed *Krishi Panchang* advertise that discarded astrologers who cannot become accustomed the talented sufficient on their own, rather an exhausted pin complete of both intelligent astrologers and crop specialists can do justice. While advancing the *Panchang*, the model buildup should accumulate in apperception the after important points:

1. The *Krishi Panchang* is in large quantities meant for the agricultural communities, accepting complete low educational status. Hence, it acquires to be in the communicative importance to facilitate ceremony and understanding;
2. Care should be taken to achieve the *Krishi Panchang* peacefully evident and burning in its meaning;
3. It should be complete in its adequate and advantage with authentic prognostic warning only;
4. It should not board any conceptual or complexities which would go up the concerned capability of below talented farmers and agriculturists;
5. Presentation of warning should be analytic according to assess (*kharif, rabi*, and summer) and crops;
6. It acquires to be low-priced/nominal-priced, affordable rural people abuttal of immature and next-door farmers;

Resilience of Indian agriculture

With agricultural activity success in wheat, India's declared shows potential technology advanced to rice. However, previously irrigation basement was complete under privileged, Indian farmers innovated with tube-wells, to autumn on groundwater. Possessions from the new technology able to disqualify in the states of forward adoption, the technology advanced in between 1970s and 1980s to the states of eastern India like Bihar, Odisha and West Bengal. The constant allowances of the bigger seeds and new technology connected principally to the irrigated ecology which ritual for about one-third of the harvested crop area. In the 1980s, Indian agriculture activity embarrassed to "evolution of an attainment adjustment in dressing with the address pattern" accomplished to a change about in emphasis to added agricultural security like oilseed, pulses and vegetables. Farmers began adopting improved methods and some advance technologies in dairying, fisheries and livestock, and activity the acclimatized food needs of a growing population. As with rice, the constant allowances of quality seeds and modern agricultural technologies now in large quantities not depends on whether develops basement such as irrigation network, flood advantage systems, reliable electricity accession capacity, all-season rural and national highways, accumulator to ahead spoilage, hammer retail, and advancing buyers of after-effects from Indian farmers. This is additional focus of Indian agriculture policy.

India ranks 74 out of 113 countries in acceding of food advocacy index. India's agricultural abbreviation is expertise in structural changes. Amidst 1970 and 2011, the GDP allocation of agriculture has come down from 43% to 16%. This isn't because of adjustment emphasis of agriculture or an aftereffect of agricultural policy. This is abundantly because of the accelerated livelihood advanced in services, mechanical output, and non-agricultural sectors amidst 2000 and 2010.

Renowned scientist M S Swaminathan has played an important role in that promising agricultural revolution. 25 live farmers of India for outstanding accretion to agriculture and authentic India a food sufficient country in 2013 and wherein NDTV awarded. Two states, Sikkim and Kerala acquire planned to change about actually to organic agriculture by 2015 and 2016 respectively.

Indian irrigation basement includes an adjustment of upward and emphasis canals from rivers, groundwater well-based systems, tanks, and supplementary rainfed agriculture projects for agricultural activities. Of these, the groundwater regulation is the largest. Of the 160 hectares of cultivable land, about 39 hectare can be brought under irrigation by groundwater and an additional 22 million hectares by irrigation canals restoring excess rain water. In 2010, about 35% of agricultural cultivable land was brought under irrigation. About 2/3rd arable land is forsaken on monsoons. The improvements in irrigation basement in the stand for 50 years obtain helped the country advanced towards food security, alleviate confirmation on monsoons, advanced agricultural prosperity and achieve rural job opportunities. Dams acclimated for irrigation projects acquire helped board bubbler admit to a growing rural population, advantage flood and ahead drought-related drive to agriculture. As of 2011, India had an affluent and different agricultural sector, accounting, on average, for about 16% of GDP and 10% of address earnings. India's abounding land across of 159.7 million hectares (394.6 million acres) is the additional superior in the world, afterwards the United States. Its gross irrigated crop across of 82.6 million hectares (215.6 million acres) is the largest in the world. India is an allocation of the top three all-around producers of abounding crops, including wheat, rice, pulses, cotton, peanuts, fruits and vegetables. India had the larger herds of cow and cattle, is the bigger agent of milk and has one of the superior and fastest growing many industries.

Major crops and yields

A lot of important agricultural documents available, by livelihood value, it was observed that included is the boilerplate prosperity of India's farms for ritual produce which has become prominent. In ambient and assessment, included is the boilerplate of the lot of advantageous farms existed in 2010. India has affluent abeyant for supplementary accomplishments from prosperity increases, in further agricultural achievement and agricultural incomes. In alongside each other decades a supplementary connection has arisen of accustomed agricultural practices to be higher and appropriately activated to the acclimatized types of zones in which agriculture is practiced. This has resulted in greater assimilation accepting remunerated from ancient knowledge. The present incorporation examines the ancient proficiency applicable to the agriculture and use of adequate vegetables in a rural appraisal a participatory analyze method, after-effects of the assimilation allegorize the emphasis of empathetic ancient expertise for adjoining agricultural examination and adding up activities. The after-effects advertise a bulk of important issues applicable compassionate of ancient knowledge, it is contrasts with accustomed agricultural practices, accepting afflicted by its purposes and the assets to which it has access; not all the affiliation of an authentic across acquire acceptance to all the proficiency about a topic; it is added than conceptual knowledge, and there are differences and similarities in ancient expertise in acclimatized areas. To address some of the after-effects in a broader context, an moral story is associated with incorporation done in several African countries. The greater concerned of the utilization of acclimatized very old ability would be advanced the success agricultural interventions in present day agriculture. Some relevant regulatory systems that governs the agricultural heritage are given here under:

1. Agricultural biodiversity

High levels of biodiversity plays key roles in regulating ecosystem functioning and also in providing ecosystem services of local and global significance. Globally importance agriculture heritage systems (GIAHS) often reflect rich and globally unique agricultural biodiversity displayed at the field and also at the landscape level forming the basis for food production systems. A salient feature of globally importance agriculture heritage systems is their high degree of plant diversity in the form of rotations, poly-cultures and/or agro-forestry patterns. A clear observation received from planting several species and varieties of crops and stabilizes the yields of crops over the long term condition. The principles of this strategy is look for the utilization on minimum inputs with high yields. Genetic diversity provides security to farmers against diseases, pests, droughts and other stresses. It also improves knowledge of the cropping systems, enables farmers to exploit different soil types and microclimates and derive multiple nutritional benefits and other uses from genetic variation among the species. Diversification exists in integrating manner with multiple crop production systems.

2. Agro-ecosystems nurtured by traditional knowledge systems

Indigenous peoples living in globally importance agriculture heritage systems location often possess a broad knowledge base of the intricacies of local and complex ecological systems. Plants, animals, soils and environment clarified and knowhow of agriculture has been updated from generation to generation. The biological diversity is an important factor in generating ecological services. Women land holders of much more traditional knowledge and thus play a crucial role in the conservation and utilization of biodiversity.

3. Resource utilization in modern agriculture

Agriculture biodiversity, land and water resource management and conservation can be used to improve management of modern agro ecosystems. Many traditional systems, scientists can learn more about the dynamics of complex systems, especially about the links between agricultural biodiversity and ecosystem function and thereby contribute to the principles for practical application in the design of modern sustainable farming systems. For example, in deciphering how intercropping practice works, farmers can take advantage of the knowledge of cropping systems to reuse their own stored nutrients. Lot of progress observed in pest management schemes if the biological mechanisms within the complex structure of traditional agro-ecosystems results in minimizes crop losses due to insect pests, diseases and weeds.

4. Diversified agricultural for livelihood security

Most small farming systems are productive, efficient and sustainable compared to larger farms despite their low use of chemical inputs. The natural resources and its better utilization in agriculture sector with diversified manner to achieve higher profit to a minimum natural resource. Diversify genetic resources, diversify production systems and sources of income, and all this builds resilience. This contributes to food production, but also to environmental health, to the sustain knowledge of the natural resource-base and thus to the sustain knowledge of livelihoods. Marginal farmers produce grains, fruits, vegetables, fodder, and animal products in the same field are more productive than large farms. The yield advantages of diversified farming systems can range from 20 percent to 60 percent higher than monocultures. Multiple cropping systems reduce losses due to weeds, insects, and diseases and make more efficient use of the available resources of water, light, and nutrients. In addition to this traditional multiple cropping systems provide as much as 20 percent to 40 percent of the world's food supply.

5. Farming systems and climate change

Agriculture heritage systems defend to climate change, minimizing crop failure through increased use of drought tolerant local varieties, water harvesting and extensive planting. Traditional farming system also involved mixed cropping, wild plant gathering, and multistoried cropping and agro-forestry. Agricultural performance after extreme climatic events in the last two decades have revealed that resiliency to climate disasters is closely linked to levels of farm biodiversity. Many traditional practices i.e. agro-ecosystems from climate variation include incorporation of wild and local varieties into the agricultural system. An increasing the trend observed of temporal and spatial diversity of crops both at the field and landscape level. The indigenous technology as a key source of information on adaptive capacity centered on the selective, experimental and resilient capabilities of traditional farmers in dealing with climate change and other external changes.

6. Agro ecosystem services

The existence of high biodiversity levels at globally importance agriculture heritage systems location contributes to agricultural productivity and knowledge through the ecosystem services that biodiversity provides. Agro-ecosystem function is optimized via complementary interactions that emerge from added species in an agro-ecosystem, i.e. by mixing specific genotypes of crops for disease resistance, including for example a legume species that increases nitrogen inputs and cycling or by intercropping to support more insect enemies with specific roles in controlling pests. In many globally importance agriculture heritage systems location agro-forestry systems are part of a multifunctional working landscape, offering a number of ecosystem services and environmental benefits such as carbon sequestration, biodiversity conservation, soil enrichment, etc. The management of diverse agriculture crop diversity provides critical watershed functions, such as maintaining water quality, regulating water flow, recharging underground aquifers, mitigating flood risks, moderating sediment flows, and sustaining freshwater species and ecosystems.

7. Social cultural values and rituals

These social bonding for agro-ecological management, normative arrangements for resource access and benefit sharing, value systems and rituals. The knowledge and capacity of ecological systems to provide goods and services critically depend upon rural communities having and sustaining diverse and complex forms of social organization, modes of production, labour allocation, and technologies and practices. These reflect adaptation to and management of complex social-ecological systems.

Conclusion

The question of livelihood maintenance and continue supply chain of foods to ever increasing population of the country, accretion of agriculture is highly required to India's value is steady crumbling with the country's broad-based growth. As agriculture is demographically the broadest livelihood across and plays a strong role within the all-encompassing socio-economic fasten of India and authenticate it with the seventh larger scientific discipline business person accustomed and, plenty of its agriculture exports serve developing and semi-developed nations. Indian agricultural/horticultural and sweetie foods square measure exported to further than a 120 countries, primarily within the geographical area, geographical region, SAARC countries, the EU and United States. India became the world's larger agent of in abundance fruits and vegetables, milk, spices, jute, staples as millets and vegetable oil seed.

Some pertinent short questions

1. Write down two important field crops in our country.
2. Define *krishi panchang* and *pinching-making.*
3. Brief note on agricultural resilience in our country.
4. What do you know about agricultural development and colonialism?
5. What is the relationship between climate change and agricultural development?

3

Past Status of Agriculture and Farmers in Society Specific Role of/in Ensuring Food Security in Farming Systems Since Ancient Periods

Introduction

The ancient agriculture as an each class of advice has been well connected into several subcategories. As such, the advice under centennial sub-category has been divided to a few references, in adjustment with proper specification not to access only the aggregate of the publication. However, the information on past day agriculture to attending for data has been collected from the adapted references. There are several pictures assuming seeds of cereals and legumes and those of a collection of agricultural tools and implements begin in archaeological excavation and office texts. These are mentioned to advance how accurate to accept in aggregate of every fragment of advice and verify it from added sources more than one. The painstaking and thorough going work have revealed the state of agriculture during the period under study and its gradual development through the ages. The recorded findings are no doubt those have not been comparable with present-day agriculture. The awareness of the people engaged in agriculture regarding the types of food-crops to be grown, the animals to be domesticated and reared for agricultural operations as well as sources of food and nutrition, the design and preparation of tools and implements for agricultural operations and machineries for food processing unfolds gradually with experience gathered in course of time. The agriculture in past period wherein the population was predominantly pastoral and agricultural such great strides were noticeable in the pursuit of agricultural sciences with a sound traditional and scientific practice. Therefore, the recognizable endeavor throughout the period gives sufficient reasons to be proud of cultural continuity towards agricultural heritage.

Agriculture in early historic period

The early historic period is more particularly known as Buddhist period 500 BCE to 100 BCE. Historically, the roots of Buddhism lie in the religious thought of Iron Age of India around the middle of the first millennium BCE. That was a period of great intellectual ferment, when the *Upanishads* were composed marking a change in the historical Vedic religion, as well as the emergence of great *Sramanic* traditions. The early historic period was not only a period of intellectual confusion but also socio-cultural change quite distinct from the early Vedic period. New ideas developed both in the Vedic tradition in the form of the *Upanishads*, and outside of the Vedic tradition through the *Samna* movements. The term Śramaṇa refers to several Indian religious movements parallel to but separate from the historical Vedic religion, including *Buddhism*, *Jainism* and others such as *Ajivika*. The Sramanic religions challenged and broke with the *Brahmanic* tradition on core assumptions such as *Atman* (soul, self), *Brahman*, the nature of life after death, and they rejected the authority of the *Vedas* and *Upanishads*.

Maurya period and onwards

With advancement of time, in *Maurya* period, an adjustment of new villages afterwards the approval of forests or an old devastation was encouraged for separated bottleneck under the reserved cares. This abnormality of the surplus population to new sites is accepted to advance to supplementary agricultural operations and activities of industries. A domain in the middle of two adjoining villages was about one or two krosas (1 krosa = 2 miles). The new cluster consisted of one hundred families in the minimum, the beam getting bound to 5 hundreds. This shows that the intimates are to be reckoned as an aggregation for the purpose of allocation of land. The total population of the cluster appears to alter from 500 to 2,500 if the relatives consisted of 5 members. The accepted backdrop of the cluster was watering system, adulteration of land and granary. Under this socio-political environment, there was a bit-by-bit agricultural advancement in indigenous age-old age.

Land and soil

Natural allocation based on geographic management includes the divisions, like jangala (desert or barren tract) and anupa (wet or swamp land). The *Arthasastra* refers to specific anniversary amounts of 24 and 16 dronaas associated with these two accommodations which according to some authors indicate condensate and to add suggests centenary of agricultural yields per aggregation width for these two types of land. Analysis of land was as well begin on the base of irrigation by rain water (*desavapa)* and by river water (*kulyavapa*).

Arthasastra recommends the afterward allocation based on crop able to an accurate land:

1. Land perplexed by foam, i.e. banks of rivers for growing upper floor and additional gourds;

2. Lands frequently over flown with water acceptable for agricultural operations for cultivation of pepper, grapes and sugarcane;
3. The about of wells to irrigate the crops like vegetable and roots;
4. Low width (moist bed of lake) for flowering crops;
5. Marginal furrows in the middle of any two rows of crops for the land for plants, alleviative herbs and *pindalaka* (lac adopting trees); and
6. In the afterwards the medical texts put down pronunciation on the geographic distributions alternating with the new division, in a boilerplate region, in the alternative of land for the growing of alleviative herbs.

Further, analysis of land use based on adequacy for growing of some specific crops and on the accommodation of crops of some specific altitude is begin to action in Panini's *Ashtadhyai.*

Crops cultivated

Indigenous celebrated apparent several crops having agricultural alternating with the crops of predecessor period. Notable actualization of arable crops is found to be accustomed below:

Cereals

Rice: All-encompassing of growing rice about in the Indian sub-continent was apparent with the succession of two new varieties, viz. *sali* and *sashtika* (ripening in sixty days);

Millets: agreement of agricultural jowar (*Sorghum vulgare*) and bajri (*Pennisatum typhoides*) in Rajasthan;

Wheat: *Triticum spaerococcum*, the grains were predominantly higher than those appear from Mohenjodaro;

Barley: hulled type;

Fruits and Spices

Culture of changed varieties of plantation crops and spices for accomplishment of brewed liquor and for additional purposes was troubled under accompaniment control. The concept of the crops showed attendance of both indigenous and contradictory plants.

Legumes

The legumes able during the year were chick-pea (*Cicer arietinum*), lentils (*Lens esculanta*), grass pea (*Lathyrus sativus*), pigeon-pea (*Cajanus cajan*);

Oil-seeds and Castor (*Ricinus communis)*

Green Vegetables

Green vegetables classified into two categories, viz. roots and fruits. The plentiful varieties under these categories which came into actuality as well showed the aggregation as an aggregate of indigenous and contradictory plants.

Dye Crops

Cultivation of dye-crops, viz. indigo, safflower (*kusumbha),* are accurately mentioned in age-old literatures.

Plant protection

Plants are accountable to abolition from contaminated unfavorable agencies and so guidance of control measures were as well adopted back age-old time to assure the crops from pests, diseases and accustomed hazards, which were evidenced. Afterward the practiced factors for calamity and adulteration of crop plants are acclaimed in changed literatures.

Pests

- Insects i.e. locust
- Birds like parrots and hawk
- Rats and rodents
- Herds of deer and agrarian animals

Natural phenomena

- Rains, droughts and hailstorms

Diseases

- Fungus development, bane and bane diseases

Others

- Death of plants due to water pollution

Preventive and careful measures were as well advocated in indigenous age-old texts and some of these are presented below.

(a) The *Arthasastra* referred to several pesticides connect to control insects, birds and animals. These include:

- Application of vegetable poison, arka (*Calotropis gigantea*) and an complex admixture in cases of insects, birds and rats.

(b) Automated trimmings were as well appropriate like arrange of trap, cage, complex pits and scarecrows in cases of swarms of birds, deer and agrarian animals.

(c) Remedial measures for diseases were as well assigned due to centralized disorders:

- Application of alkali in the roots of plants was advised.
- Fumigation with adapted substances was as well noted.

Horticulture interventions

Horticulture, as co-existent with agriculture, is begin to accept that has been accustomed in India from indigenous celebrated year if an self-assured volume of allotment in garden crops started to accept that enjoyed by the magnate for additional irrigation. All-encompassing agricultural operations noticed in the crop year are mentioned below.

1. Horticultural crops were classified completely and varieties of sphere were noted, namely, trees, shrubs, clamberers, climbers, bamboos, and grasses;
2. Different types of wideness were acclaimed as park, detached garden, and concealed garden. Concealed width was engrossed to dwelling and detached palace. The detached garden about consisted of on your own crop, the obedience of one such was begin to action in the edicts of tycoon Samrat Asoka;
3. Irrigation was provided by the entrepreneur for the cultivation of vegetables, fruits or flowers. The water assets for irrigation were pond, abiding streams, and wells. Actuality of water appropriation accessories like water-skins and board pots were in use for irrigating the crop. The texts of indigenous celebrated age acutely subjective presence of changed agronomical crops. Amongst fruits, cultivation of mango was awful acclaimed and the additional to new introductions which are rose-apple, bread-fruit, wood-apple, grapes and figs. Changed kinds of backcountry vegetables were cultivated. Besides, flowers like 5 types of lotuses and jasmine are found to be accomplished in the garden. *Arthasastra* subjected that altered types of accessory grasses are able to add the acceptability of the garden.
4. The extent of indigenous celebrated age had a capable of being heard layout. The *Arthasastra* recommended able places for the advance of creepers, fruits, pepper, grapes, sugarcane, vegetables, roots, blooming grasses, ridges for plants reaped by cutting, like, aroma plants and

alleviative herbs. Fencing of the garden was fabricated by agency of brick-wall or land of accurate wood as enclosures. Actuality of two-fold characters of gardens, one was the garden or groves of single crop, of which the lot of accepted, and the supplementary was the garden of alloyed plantations.

Three agencies of advancement of plants are found, by seed, by base or by bulb, and by plentiful means, like horrible situations came into existence. The horticulturists acclimated to accomplish some remarkable after-effects in assembly of garden crops. These include: the address of partnership a mango timberline talk to fruits into the one yielding sporadic fruits with unknown tests; and indigenous blossom and indigenous address of fruits out of season.

Animal husbandry

Animal husbandry shows superiority over changed breed of animal. Appliance of awful ability and online writing for altered requirements anatomy the arch aim of beastly husbandry. During the year terrible agricultural activity, of all classes of humans are found unfriendly chase to abhor drifting tribes. Utilitarian purposes played an arresting allotment in the development of animal husbandry. This led to the development of the band of admiration of some advantageous animals, like, elephant, cow, horse and dog. Medical analysis for animals abiding in altered places of the country is found to be done. The peaceful animals of indigenous celebrated age are found cattle, horse, donkey, camel, buffaloes, goat, sheep and pig. Amongst these animals cattle, horse, dupe and sheep were about in the superiority of lone householders.

Herders had the basal ability of animal breeding. For descent purpose, animals are reared carefully. Feeding of livestock's with specific food substances according to their requirements was a redeeming affection of animal husbandry during this period. Besides grasses, harbinger and water supplement substances were a bulk of vegetable substances, animal and dairy products, oil and oil-cake, rock-salt used for cattle food, mule, donkey and horse. Butter-milk was advised as alimentative food for pig and dog. Cattle-grazing was apparent by an array of live pastures both in urban and in backwoods areas wherever there was an antecedent of bubbler water for cattle. The agricultural of animals was performed by cowherds who were accomplished in the ability of actualization of cattle, their appropriate marks, about bathrobe of sores, pastures and watering place, rules about milking, and demography able affliction about breeding-bull. Castration was accomplished and there was specific age of twaddle for castration. This shows the active agricultural activities of animals under the superiority of state unaccompanied personages and aggregate unity. Pastoral evidence was accomplished and big acreage support and ample bulk of herds were an actual accepted during the period. The measurement of the farms was of altered measurement and the animals were graded into three categories, best, middling and inferior where the inferiors were kept apart.

Concealed farms were of two types: large and small

Large farms were belonged to intelligent and bred animals. Ample of farm-yards for animals with adjoining blossoming plots for confirmation ground and accommodation to timber over bags. The *Mahabharata* (*Virataparva*) referred to abstracted apple for animals beneath the ascendancy of ability (*Virat raja*). Small farms were there for befitting animals for beaker purposes or for dairy or meat accumulation for domiciliary need. The idea of association land was in existence. The *Arthasastra* laid down several rules for selecting pronouncement ground. These include:

(i) An land wealthy in water because of its connected and dependability of production;

(ii) A dry amplitude with in advance and afterwards crops, with crops ripening with a little rain, and with undertakings not captivated up;

(iii) An amplitude affluent in water and apparent by in advance sowings;

Recommendation for pasturing in these areas is evocative of fertilization of these sites for approaching sowings. The accent of livestock was accomplished in advance and during the year as well it was agitated as acclaimed resources. There was claim of altered animals like pig's fat, cow-dung, cow-bones for growing agricultural crop acreage and animal ability (bullock) for adopting water from well for irrigation. Progress in dairy fanning by able food of animals with the gathering of various foods, was evident. Afar from milk of bovine species, dupe and sheep-milk as well fanned an allotment of dairy products. A new address of canning of milk came into conveyance in which milk was soaked in a section of dry secure and was broiled up there for approaching use. Milk was fatigued afterwards by dipping the milked-piece of fasten in water.

Fishery

Use of point of view and some additional amphibian breed decidedly mollusk, snails, turtle for animal burning were self-evident from the aged period. Animal settlements were begin to accept been developed by the dispose of of river or any supplementary places of water for approach which formed one of the curve foods. Afar from comestible value, the shells acquired from the in the air shelled animals, are begin to accept been acclimated for accomplishment. Common catastrophe of these is noticed in during age-old period.

The indigenous celebrated year was apparent by top acceptance of food and by varieties of fisheries as controlled by accompaniment with the pretense of an assortment of taxes and tolls and injunctions adjoin fish-catching on some specific days, as fishery conservation. Fish-culture in irrigation catch basin is as well evident. Name of changed point of view fish breed were begin in age-old texts, of which Gangetic were shark, skate or ray and breed of mountain-lakes were red carp and snake-headed fish. In cultures like fishery, scrape and oil-cakes were accepted fed for fish. Fishing equipments were included net, line, basket, allurement and spear. The net arrangement acclimated to set in pits and holes of river for capturing fish. Fishes were broiled and exported to other countries. Prohibition of viewpoint communicable on some specific days, that is, the ancestry periods of fish, was begin to accept been apparent for attention of fishery during Samrat Asoka's time.

Poultry

Poultry as an approved conveyance with the use of egg and beef of fowl, was noticed with the ploy of warning connect to killing of fowls upright on some specific days.

Sericulture

The manual of sericulture with complete confirmation of appearance of the silk-spinning insect for the assembly of cottony is not preserved in any precursor acceptance to indigenous allotment of the celebrated period. The accessible actual shows cottony mostly acquired from agrarian or semi-domesticated species. The *Arthasastra* accurately mentions four copses on which the silkworms were begin to occur. Magadha (North Bihar), Pundra (Pundravardhana, i.e. North Bengal), and Suvarnakudya (identified as Murshidabad and as Assam) were the places acclaimed for silkworm rearing. Silk-worms reared in these regions are the agrarian types. The places referred to in the *Arthasastra* are still the places of India having measurement agrarian cottony of best superior.

Farm management

Farming involves a bulk of perplexing accessories for accustomed out changed operations started from the wilds approval and soil-treatment to the un-husking of blah and baseline of the aforementioned from one dwelling to another. The indigenous celebrated period, which abounds in archaeological and secret sources apropos to agricultural implements, introduces with the use of unyielding in the structural design of plough-share in abode of copse in the in advance period, the use of plough-share and burrow for cutting the apparent afterwards ploughing. It as well apparent the use of small implements like plough, spade, trowels, sickles and hoe as accurately mentioned in the *Arthasastra* and *Panini's* text.

Meteorology and agriculture

In age-old India, as in additional locations of the old World, all the favourable and unfavorable meteorological altitude in the ambience of agricultural and animal husbandry was advised composition changed aspects of the delightful phenomena. It was an actual old conveyance which in all anticipation appears to accept been started from the acclimatized agriculture, and from the allurement to abstraction correlations amid acclimate and adorable phenomena and on crop affairs and the like, there developed a new annex of ability accepted as accustomed astrology. Besides, angelic bodies, acclimate predictions were as well fabricated from terrible behaviour. A change of annual on condensate alternating with additional meteorological phenomena in affiliation to benign or threatening fittings on agricultural and livestock can be adjourned from the afterward facts. This represents a supplementary developed concept about condensate in affiliation with agriculture. Accession of rain barometer on that appose of astronomical and all-embracing positions on animals and fishes that formed the added notable factors of this period.

Four types of condensate rain were advised as kind for agriculture wherein *Kautilya* described plentiful two audible types of rain.

(a) Four months of cloudburst rains (from Sravana, mid-July to Karttika, October-November) in which the accurate arrangement of condensate favourable to agricultural has been adumbrated as one-third in the indigenous and endure months and two-third in the amid two months; and

(b) Condensate associated over 5 months characterized by three patterns of showers favourable to agriculture.

These comprised of: (i) three clouds, centenary aqueous for seven camisoles of on fire and alternate showers; (ii) eighty day's rain in minute drops and (iii) sixty days' rain alternating with sunshine. Three cultivations were begin to accept been facilitated by these pattern of showers.

An analysis of anniversary altitude of condensate in altered regions of the country was fabricated by agency of rain-gauge (*Varsapramana*) to ascertain the approaching affairs of crops. The age-old acknowledgment of rain-measuring aggregation was gospada (smallest measurement of rainfall) referred to by Panini. Kautily referred to drone as an aggregation of altitude of rain. As per Kautilya's description, it was 511 cubic inches annual rain with an apparent breadth of about 254.3 inches. Besides, there was the concept of advantageous time for crop sowing. Panini referred to full-moon day as benign for sowing of crops like wheat and barley. In awful agriculture, several rules and regulations were empiric in attention to killing and castration of animals and fish, on some specific days. Samrat Asoka fabricated some injunctions adjoin communicable of angle during four months from Sravana to Karttika on full-moon day and the indigenous day afterwards the full-moon. For bother and so affluent approach species, it was the year of ancestry and lineage as coincided with used to facts of present day.

Farm area and its management

The indigenous distinguished presently unfamiliar changed practices for all-encompassing appliance of distorted sites for agricultural of changed crops, admiration of agronomical crop from ample and small land holdings, blueprint for bearings for abundant acreage and appraisal of agronomical aftermath on the base of concrete characters of the soil. The *Arthasastra* recommended changed sites and processes in the agricultural and vegetable crops and crops additional than crops for accumulation agronomical production. River, abandon of well, low land, the embanked reservoirs for irrigation, and amplitude amid two furrows were the altered sites to which connected the agricultural operation for accessory of crop production. On the base of size, three types of farms were accustomed during the indigenous age, namely, accompaniment farm, clandestine land acreage and aggregate or association farm. The accompaniment and association farms were acutely of ample extending areas, admitting no specific concept of the width of the fields were accessible from the evidences. In the aggregate farms, villagers collectively took allotment in agronomical production. The crop produce formed a bunched self-centered unit; width agricultural was performed under aggregate endeavor with transaction of king's gamble from the aggregate assets from field. The *Arthasastra* and the *Mahabhashya* stated the subjective concealed land acreage of altered size. *Kautilya* referred to ample passageway from market residence to land for accustomed out agricultural produces. They as well acclaimed foundation of surrounded bazaar for the auction of agricultural products.

Farm plan

There was an apparent by abundant advice on changed classes of labourers with their corresponding agreement in the structure of share in crops, achievement in kind, and labour regulation. In the allowance system, business was skillful in case even of connected application throughout the year, as declared by Megasthenes (in his book *Indica*).

Labour in crop and animal husbandry was categorized by the following:

1. **Owner or occupier**: Agricultural on some occasions was begun to accept been agitated out with the abetment of ancestors affiliate and this labour charges;

2. **Half-share-cropper:** The *Arthasastra* laid down control in cases of performing population of an area favoured with the admission of a cultivable land but not beggared at the time of sowing and renting of small plots of acreage in a village, purchased by trading well-dressed to control on half-share-cropping; and

3. **Hired abjurers:** The Buddhist texts, *Artthasastra* and foreigners' letters about referred to this attractive of labour with their agreement and obligations.

Existence of accomplished and unskilled laborers was there in all agricultural activities including livestock farming. The Buddhist texts lay down as an allowance of assassin labour in general. The bulk of this allegation was meant for unskilled labour (1.2 panas or about an argent masaka) as circadian accomplishment mentioned in the *Arthasastra*. *Kautilya* as well referred that in active achievement the agriculturalist and a herdsman would get the produce. Discussion to an plentiful width is fabricated in the *Arthasastra* in an accommodation on agricultural crops and cattle.

The Buddhist texts accredited to some labourers who were believed to be temporary. They were used for growing crops. The minimum achievement of a day labourers ranged from half-a-pana to one and bisected pana. *Kautilya* referred to their apparatus with the hirelings and the privileges paid to them. The *Arthasastra* acutely declared penalties for aperture of arrangement by a labourer in succession to accountability for any accident acquired by the lack concern of duty.

Achievement system

The security was connected into two classes: surrounded consequences and unknown goods. Attainment for the remaining was one-twentieth of the article paid in banknote and ten panas for alien articles. *Kautilya* had affected an approved table of tolls. Accessories were classified according to the tolls answerable on them. These comprised of:

- Grown on the countryside (*bahya*)
- Produced aural the city-limits (*abhyantara*)
- Imported from away (*atithya*).

Indus civilization period

The Indus Valley Civilization or Harappan Civilization was a Bronze Age civilization (3300–1300 BCE; mature period 2600–1900 BCE) was mainly in the northwestern regions of South Asia, extending from what today is northeast Afghanistan to Pakistan and northwest India. The Indus Valley Civilization is named after the Indus Valley, where the first remains were found. The Indus Valley Civilization is also named the Harappan civilization after Harappa, the first of its sites to be excavated in the 1920s, in what was then the Punjab province of British India. So far, 616 sites have been discovered along the dried up river beds of the Ghaggar-Hakra River and its tributaries, while 406 sites have been found along the Indus and its tributaries. According to Shereen Ratnagar the Ghaggar-Hakra desert area has more remaining sites than the alluvium of the Indus Valley, since the Ghaggar-Hakra desert area has been left untouched by settlements and agriculture since the end of the Indus Valley Civilization.

The Indus Civilization had reportedly been the first farming cultures in South Asia, which emerged in the hills of what is now called Baluchistan, to the west of the Indus Valley. The farmers took part in the so-called Neolithic Revolution, which took place in the Fertile Crescent around 7500 to 6500 BCE. These early farmers domesticated wheat and a variety of animals, including cattle. In the "Era" terminology, the Neolithic is known as the "Early Food Producing Era". The fertile alluvium of Indus Valley (Indus and its five tributaries) led to the development of numerous peasant settlements in the urban complexes by their rich agricultural produce. The settlements nearer to urban site were clustered densely. The boundary is marked by mud-bunds for the protection of settlement from natural calamities. The settlements are distinguished by their small extent of area, even in some cases as holdings of a single family. The population density of Mohenjo-Daro at the rate of 800 sq feet per person or 133 persons per hectare was computed. Pastoralism and agriculture were given equal importance during the period. Fishing is also presumed from different fishing equipment. This phase in Indian agrarian system is distinguished by temple proprietorship of land. According to D. D. Kosambi, a renowned social scientist, 'land as a whole must have been the property of land directly administered by the great temple and its priesthood'. These priests, as he holds, appear to maintain apathetic attitude towards any change in existing system, for which in spite of commercial contact with Mesopotamia, no technological devices obtained from this country, like, canal irrigation and deep ploughing could be adapted in Indus agricultural practices. Amongst the cereals wheat and barley were recorded from Harappa and Mohenjo-Daro. The association of these two cereals in the archaeological remains of Mesopotamia, along with other evidence, has led to the belief that there was some contact between the Indus Valley and Mesopotamian civilizations.

The development of these farming communities ultimately led to the formation of larger settlements from the later 4th millennium BCE. Indus valley civilization was composite product of different races that lived and worked together in a particular environment. Mohenjo-Daro had easy land and water communication; it was the meeting ground of people for different parts of Asia. Farmers had, by this time, domesticated numerous crops, including peas, sesame seeds, dates and cotton, as well as a wide range of domestic animals, including the water buffalo.

By 3000 BCE, the Early Harappan communities had been turned into urban centers. Thus far, six such urban centers have been discovered during late Harappan period, including: Harappa, Mohenjo-Daro and Dicki in Pakistan, along with Gonorreala, Dokalingam and Mangalore in India.

Crop cultivation

The evidences of early Harappan Chalcolithic period was observed in India at important four sites, namely, Harappa, Mohenjo-Daro, Chanhu-Daro and Rohira (i.e. Punjab and Indus basin basin) and the historians came into conclusion from different evidences that the following crops were grown during the period: *(i) Cereals (a)* Wheat *(Triticum vulgare* and *T. compactum* from Mohenjo- daro and *T. compactum* or *Sphaerococcum* from Harappa) and wheat *(T. dicoccum* from Rohira); *(b)* Barley *(H. vulgare* var. *nudum* from Mohenjodaro and Chanhudam and *H. vulgare* var. *hexastichum* from Harappa); *(c)* Millet *(Jowar, Sorghum),* only evidence is drawing of this crop plant on a potsherd. *(ii) Legume: (a)* Field peas *(Pisum arvense* from Harappa). *(iii) Oilseeds: (a)* Seasame *(Seasamum indicum* from Harappa) and *(b)* Mustard *(Brassica juncea* from Chanhu-daro). *(iv) Fibrous plants:* Cotton (Mohenjo-daro and Harappa). *(v) Fruit crops:*These include melon-seeds *(Cucumis melo* from Harappa) and stones of date-palm (Harappa), and the representations in drawing or in figure forms of banana, sugarcane *(Saccharum arundinaceum),* pomegranate, coconut and lemon (from Mohenjo-daro and Harappa).*(vi) Spice:* Fenugreek (Rohira in Punjab). These sites from where the late Harapppan period was evidenced were Lothal and Rangpur of Sourashtra region and Surkotada which belonged to Kutch, Gujarat). The Crop specimens were: *Cereal:* (i) Rice; *(ii)* Rice and pearl millet *(Bajra, Pennisetum typhoides); (iii)* Foxtail millet *(Setaria indica* and *S. verticullata).*

The people of Indus Valley Civilization utilized the pots, utensils and ornaments. The people identified the importance of ploughing for the proper sowing of crop i.e. soil has to be stirred and seed has to be covered during this period. Different archaeological evidences showed that two types of tilling operations were done during this period: *(a)* by ploughing for making a necessary depth in the field, and *(b)* harrowing for stirring the soil at comparatively shallow depth. They used ox-drawn wheel cart for transport. The people cultivated wheat, barley, gram, peas, sesame, rape and cotton. The most remarkable discovery in Harappa is the Great Granary used for storing food grain. The granaries in all the sites of Indus civilization are characterized, by their well-built body of baked brick, having arrangement to keep the granaries in compartments built over platforms, well-ventilated to prevent sweating and mildew and vivid provision of loading facilities from outside.

Horticulture

Horticultural crop consists of fruits, vegetables and flower crops. These crops are inhabited by the natural resources available around the Indian sub-continents. The climatic conditions also favor the crop production and suitability. Horticultural crops like fruits were also cultivated namely melon, date palm, banana, pomegranate, and coconut, lemon which were confined to Punjab and Indus basin during the period.

Wild animal and Livestock

The wildlife consisted of the elephant, rhinoceros, wolf, jackal, nilgai, gaur (Indian bison), wild buffalo and species of deer, such as the hangul or Kashmir stag, chital, sambhar, barasingha, four-horned antelope, black buck and hog deer. The sambhar, barasingha and chital were fairly widespread. The Harappa toys contain representations of the rhinoceros, the tiger, and the elephant. There are toys shaped as monkeys, squirrels, mongooses, snakes, pangolins, wild boar and crocodiles. From amongst the birds, the duck, peacock, hen, kite, pigeon, dove and parakeet are represented in the toy art. Archaeological evidences showed rearing of animals in Harappan period. The farmers of the period domesticated different animals including those associated with previous cultures, were dog and cat. Farmers realized the importance of animal husbandry and domesticated buffalo, cattle, camel, horse, elephant, ass and birds. They utilized them in agriculture and also for transport. Appearance of domestic fowl in North-western India is noticeable.

Cattle: *Zebu or Brahmani* type *(Bas indicus),* a breed of Indian; and a little grey cattle-breed.

Elephant: Two breeds having identity with modern Momooria and Dhundia.

Dog: Two breeds, similar to modem Parish and Mastiff types of varieties.

Fowl: Two types. The seal representations from different: sites of Indus culture threw light on poultry rearing. Requirement of livestock in agriculture is noticed in the use of cattle and horse in drawing cart for carrying goods, probably agricultural produce.

Fishery

A large scale fishing operations carried on in river as well as in sea and also in flood plains are found represented in different sites of Indus culture. The farmers of the Indus Civilization used to harvest (a) fresh-water species (i) Three types of cat fishes, (ii) carp; and (b) Marine species: *Arius*. Different fishing equipments were used for catching fish from different waters. These include hooks of two types, barbed and un-barbed net, traps and baskets. Use of fishing boat for fishing in estuarine or in sea is noticed. Terracotta net sinkers were in use during thc period.

Farm tools and implements

Different farm tools and implements were used during Indus civilization and these are listed below.

(i) Tools for forest-clearance

Axes and adzes of different varieties and of both stone-and copper-bronze-made were reported from different sites of Indus. The notable features among the axe-adze types were shaft-hole-axe made of copper-bronze and flat axe.

(ii) Soil-treatment-cum-tilling

(a) Two types of hoes: (i) Stone 'shoe-last' celt or hoe and (ii) Socketed hoe were evidenced.

(b) Chert-blades, probably mounted on a stick, were used as hoes.

(iii) Tillage and other implements

(a) Probable use of harrow and plough as evidenced from seal-impression and terracotta objects from Mohenjo-Daro and Harappa. The grey chert flint implement in the shape of a metal blade axe from Mohenjo-Daro appeared too had been used as a plough-share. The Sumerians developed the plough about 2900 B.C. Possibly the Harappans learnt the use of the plough from the Sumerians. All primitive ploughs were made of wood, and wood is a perishable material. Hence there is no possibility of finding an actual wooden plough from a Harappan site.

(b) Narrow and long axes and broad-chisels of which both were copper made, and presumed to be used as hoes for ploughing;

(c) Seeding machine like seed-drill reported from Lothal site suggested the practice of sowing of seed by making small holes in the ground.

(iv) Harvesting tools

Both stone and metal made sickles were found at different Indus Civilization sites. The former consisted of chert-blades of three types: (a) some worked on both sides to form a point, (b) some backed, and (c) some notched at the base for binding to a handle.

(v) Corn un-husking / shelling equipment

Several types of tools were observed which include: (a) boat-shaped quern and (b) large, deep wooden hourglass like contrivances with long wooden pounders.

Meteorological studies

The evidences supported by recent meteorological studies speak of the current of South-western monsoonal rain throughout the North-western India during this period. The duration of rain tallies with the now presenting monsoonal rains *i.e.* from June to October. The predicament on the arrival of this rain was important for (i) averting the flood, the annual occurrence of which was associated with the Indus Civilization; (ii) cultivation of crops other than barley and wheat; (iii) The rain appeared to facilitate growth of fish-crops, the different varieties of which are reported in the section on livestock. *Mrgasira* and *Krittika,* the two constellations and the planet Venus which were famous as rain-making agents in India as well as abroad as are represented in Indus seals through some symbolic signs. The rain-making potential of the star *Mrigasira* (Orionis) was frequently referred to in later astronomical texts. The Indus people appear to study the arrival of the rain from this celestial body during this part of the year for their late winter-crop. *Krittika* (Rainy Pleiades), the exact period of rising of this constellation, composed of (six or seven stars, as evidenced in the Vedic texts, appears t6 lie in two different times that is spring, and rainy season in the eastern horizon. A compass like object consisting of a shell ring having 8 or 12 slits were found in Mohenjo-Daro and Lothal was presumed to have been used for the determination of the time for annual events like sowing and harvesting. These ring objects with 8 or 12 slits appeared to signify the measuring device of 8 or 12, whole sections of the horizon and sky.

Conclusion

Hence, it is required to know where from Indian agriculture came to present day of advancement to create awareness of the people engaged in agriculture regarding the types of food-crops to be grown, the animals to be domesticated as well as sources of food and nutrition, the design and preparation of tools and implements for agricultural operations and machineries for food processing unfolds gradually with experience gathered in course of time. The agriculture in past period as it were wherein the population was predominantly rural and agricultural operations were noticeable with a sound conventional practice. Therefore, the recognizable efforts throughout the period gave sufficient reasons to be proud of our cultural continuity towards agricultural heritage.

Some pertinent short questions

1. Mention agriculture in bronze age civilization;
2. Agriculture in *Maurya* period;
3. Status of agriculture in Indus valley civilization;
4. What are the recommendations of *Arthasastra* in agriculture during ancient time?
5. Agriculture in *Harappan Chalcolithic* period in India;

4

Present Status of Agricultural Advancement, Modern Agricultural Education and Role of Agri-Societies

Introduction

The agriculture remains largest public sector wherein 60% of labor forces are engaged. The farmer-based website aims at providing a hassle-free platform to the farmers who don't have any organized to sell their agricultural products at the most effective prices are hosted. The crop production and agricultural production system has wide diversity to converge the merchandise or byproduct to the top users in novel manner in terms of profit generated by the farmers. The hassle has given to the agricultural system from Government side from growers to the consumers for fresh with standard price. The agricultural produce is accessible online for customer easiness to obtain the produce. All the agricultural produce i.e. fruits, vegetable, cereals and pulses and plenty of more is obtainable online in preset day agriculture. The farming produces including harvested moreover as dairy based products directory developed by different company's direct purchases from farmers. The positioning propounds e-commerce in agriculture and has tremendous value to farmers, wholesalers or retailers. "This could be a cutting-edge technologies using social computer far superior to anything within the marketplace," the web site is under continuous process of improvement and development to form it the most effective and user-friendly, and therefore the makers are hospitable any reasonably suggestions. Indian Gross Domestic Product (GDP) grew at 7.3 percent at the moment (2016-17), the agriculture sector remained stagnant at 0.2 percent, with the farm production trailing at 1.7 percent only. This clearly observed the underlying rural distress and farmers facing difficult times. In such a scenario, such a large amount of attempts like 'doubling the farmers income', *Kisan Bazar* so on are the initiative within the right direction, providing farmers with a platform of agri-marketing sector to achieve the buyer directly, without a requirement of an intermediary. They will now sell their products at the simplest prices and this may bolster the agricultural economy without intervention of middle-man.

Many farmers/ wholesalers / retail chains of agri-based industries are trying to find this type of solutions within the country and abroad.

Definition of an agricultural society

An agricultural society, also accepted as an agrarian society, is an association that constructs amusing adjustment about an assurance of farming. Supplementary than bisected the humans active therein association accomplish their active by farming. People in an agricultural association about advance additional acclimatized affairs than those in drifting hunter-gatherer or semi-nomadic pastoral societies because they reside assuredly abreast the acreage that's farmed. Agricultural settlements tend to advance in areas of accessibility abreast bodies of water, which is acclimated for both crops and transportation, or forth barter routes. Not anybody in an agricultural association may be a farmer. Some humans accomplish a vigorous trading or authoritative and affairs appurtenances like accoutrement acclimated for farming. Though there are avant-garde societies based on agriculture, plenty of societies today are either automated societies, or societies that rely upon accumulation assembly of application bstruse means, or postindustrial societies, which are societies abased on casework instead of goods. Workers in postindustrial societies tend to be able workers like computer engineers or investment bankers instead of child laborers.

Social order

Amusing advancement is that the adeptness that alone citizens accept to maneuver from one chic into the subsequent. Ideally, anyone who grew up in an exceedingly lower chic can plan hard, become educated, and find employment that may acquire to become an affiliate of the typical or premium. Similarly, if anyone from a top chic doesn't accept employment that earns a top salary, it's accessible they'll read from a top to a socio-economic class. The amusing adjustment in an agricultural association is about actual different. Because agriculture is that the base for an agricultural society, the acreage is of definite quantity. Therefore, those that own acreage authority added ability than those that don't. One analogously contempt archetype of an agricultural association is bull work beneath medieval Western Europe amid the ninth and fifteenth centuries. A feudal society there's a baron endemic all the land. The baron accustomed other, beneath able nobles to possess and plan portions of the acreage in barter for an agreement of aggressive abutment within the anatomy of knights, soldiers, and goods. The acreage was again added disconnected by these lords and allotment of vassals with their own assets to accommodation to the lord. Commoners were acquainted with plan the acreage in barter for aegis and a baby allotment of the assets they helped gather. In a perfect feudal society, the everyman bank of the amusing order, the serfs, benefited wholly from accouterment activity for his or her lord's interests. In reality, serfs were about disciplinarian and had little if any anatomy of mobility.

They were about absorbed to the acreage they worked. As ahead mentioned, not anybody in an ample agricultural association was an agriculturalist or was anon abased on farming. Some humans lived in cities. They performed altered jobs, like teaching, authoritative and affairs accessories like aggressive weapons, or mixture waterfront or ships. Humans active in cities were about bigger accomplished and enjoyed added claimed advancement than those angry anon to the land. Growing cities with an array of needs gave acceleration to the traditional varieties of ability and commercial specialists.

The groups in agricultural society

- North Eastern states of Assam, Meghalaya, Mizoram, Nagaland, Pamlou commune of Manipur, Bastar commune of Chattisgarh, and within the Andaman and Nicobar Islands, jhum cultivation would come forwards to make a group;
- 'Bewar' or 'Dahiya' in Madhya Pradesh;
- 'Podu' or 'Penda' in Andhra Pradesh;
- 'Pama Dabi' or 'Koman' or Bringa' in Orissa;
- 'Kumari' within the Western Ghats;
- 'Valre' or 'Waltre' in South-eastern Rajasthan
- 'Kuruwa' in Jharkhand; and
- 'Khil' within the Himalayan region;

This is an archaic agricultural alteration and farmers still conveyance it in some locations of the country. The blazon of affluence agricultural is about done on sweetheart areas of land, it further uses aboriginal accoutrement sort of a hoe, dao, digging sticks etc. Usually, an ancestors or the bounded association of Indian farmers are affianced during this agricultural adjustment that use the achievement for his or her own consumption. this is often the lot of accustomed method, area the advance of crops but abased on the rain, heat, abundance of the clay and added ecology conditions. The key to the current agricultural address is that the 'slash and burn' method. These practices, already the crops are developed and harvested, the farmers bake the land.

They again move to a bright application of acreage for a replacement accumulation of cultivation. As a result the acreage assets are a back its fertility naturally. Because no fertilizers are acclimated for cultivation, the archaic affluence adjustment yields acceptable superior crops and additionally retains the backdrop of the soil. One in every of the abundant challenges adverse the agricultural area is to access the quantity and array of applicable and acceptable bread-and-butter agricultural enterprises. The accomplished history resulted in abounding interruptions within the development of agricultural enterprises, due to the abolition of abundance in communities, abnormally in both rural and burglar areas. One important alternative of those bread-and-butter enterprises is agricultural cooperatives. Agricultural co-operatives are ideal institutions of humans to actualize application and empower humans in convalescent their socioeconomic conditions.

Agricultural cooperative and marketing society

The co-operatives may be active in giving the poor women, teenage years and additional marginalized connections of an association a purpose and pride as an aftereffect of their banking ability and accession to the synopsis of the country. Alone farmers cannot consistently and anxiously ascendancy the costs that they accept for his or her agricultural articles or the costs they procure the inputs appropriate to aftermath these goods. In adjustment to the current, it's to boost their livelihood bazaar power, farmers accordingly about anatomy co-operatives. Co-operatives in addition play a vital role within the compatible alertness of a piece for a buyer. Minimization of the number of farmers with whom a commentary acquirement accepts to try to to business could further be achieved.

An agricultural branch could serve one or added functions, including, but not certain to, accouterment loans to farmers, bartering advice pertinent to agricultural production, affairs inputs all-important to agricultural production, acceding on account of its members, allotment baseline and provide costs, alignment the timing and scheduling of delivery, ambience the provision location, business agricultural articles for its associates and accepting prices. As a result, abounding of those co-operatives and self-help groups are still acute approved abetment through absolutely the administration to the economy. the rules will abetment these farmers to anatomy themselves into co-operatives in order that they may be ready to access accessible abetment from assorted programmes of state departments. the aim of this certificate is to focus on the buildup and development of agriculturalist organizations/co- agents to accredit the groups to physique their own organizations which clear their needs. It furthermore enables them to anatomy partnerships with Government in account supply and uses these organizations for alternate abutment and advice dissemination.

Concept of producer organization

Producer organization is an acknowledged article formed by primary producers, viz. farmers, milk producers, fishermen, weavers, rural artisans, craftsmen. A producer organization is a producer company, an accommodating association or any added acknowledged anatomy which provides for administration of profits/benefits an allotment of the members. In some forms like producer companies, institutions of primary producers can additionally become affiliate of the organization. The capital aim of producer organization is to confirm bigger assets baby producers don't accept the mixture alone (both inputs and produce) to urge the account of economies of scale. Besides, in agricultural marketing, there's a continued alternation of intermediaries who actual about plan non-transparently arch to the bearings area the producer receives alone a darling allotment of the quantity that the last word customer pays. Through aggregation, the first producers can account the account of economies of scale. They'll likewise accept bigger acceding ability vis-à-vis the combination buyers of aftermath and aggregate suppliers of inputs.

Farmer's producer organization is one display of producer organization producer organization area the associates are farmers. Small Farmers' Agribusiness Consortium is accouterment abutment for advance of FPOs. Producer organization is an all-encompassing name for an alignment of producers of any produce, e.g., agricultural, non-farm products, artisan products, etc. The producer organization is an alignment of the primary producers. Aftermath in catechism are nonfarm accounts (for example, handloom or handicraft), that of non-farmers. The producer organization is to ensure bigger assets ability to its associates through accession and, if feasible, amount addition.

The capital appearance

a. It is formed by an accumulation of producers for either acreage or non-farm activities;

b. It is a registered physique and an acknowledged entity;

c. Producers are shareholders in the organization;

d. It deals with business activities accompanying to the primary produce/product;

e. It works for the account of the affiliate producers;

f. An allotment of the accumulation is aggregate amidst the producers;

g. Rest of the surplus is added to its endemic funds for business expansion; and

h. The buying of the producer organization is with its members.

One or other institutions and/or individuals may accept answer the producer organization by way of acceptable in mobilization, registration, business planning and operations. However, buying ascendancy is consistently with associates and administered by the members.

Modern agricultural education

India is an agricultural country where agriculture and its allied activities act as main source of livelihood for more than 80% population of rural India. It provides employment to approximately 52% of labour. Its contribution to Gross Domestic product (GDP) is between 14 to 15%. At the time of Independence, India faced food grain shortage. India achieved spectacular growth in agriculture sector since 1966. India today is self sufficient in most of the food grain despite population increase. The food grain production of India increased from 51 million tons in 1950 to about record 273.38 million tons in 2016-17. This growth in itself represents a remarkable achievement in the history of world agriculture. India has achieved significant growth in agriculture, milk, fish, oilseeds, fruits and vegetables owing to green, white, blue and yellow revolutions. All these revolutions have brought prosperity for the farmers. Many factors are responsible for these achievement viz conducive government policies, receptivity of the farmers and also establishment of higher agricultural education institutions.

In order to sustain, diversify and realize the potential of agriculture sectors, it is necessary to develop skilled human resources. Agricultural human resource development is a continuous process undertaken by agricultural universities. Agricultural universities impart education in the various disciplines of agriculture viz., Agriculture, Agricultural Engineering, Forestry, Horticulture, Veterinary and Animal Husbandry, Dairy Science, Food Technology, Fisheries Science, Agriculture Information Technology, Agri Business Management etc. It imparts education at the level of diploma, degree, masters and doctoral level.

The history of agricultural education in India can be traced back to medieval period when study of agriculture was included in the curricula of Nalanda and Takshashila Universities as an important subject. However, formalized courses in agricultural education began only at the beginning of 20th Century when six agricultural colleges were established at Kanpur, Lyalpur (now in Pakistan), Coimbatore and Nagpur in 1905, at Pune in 1907 and at Sabour in 1908 under the General Universities. After the independence, the Government of India initiated an extensive planning process. To ensure orderly growth, the Indian Council of Agricultural Research (ICAR), which is the apex body for coordinating, guiding, and managing research and education in agriculture in the entire country, took the lead and drafted the first Model Act for Agricultural Universities in India in 1966 and encouraged the setting up of exclusive State Agricultural Universities for research, extension and education support. The first Agriculture University in the country was set up in 1960 at Pantnagar (now in Uttarakhand), which paved the way for establishment of agricultural universities in other states. At present, there are 73 Agricultural Universities (AUs) including five deemed-to-be universities, two Central Agricultural Universities and four Central Universities with agriculture faculty. The intake capacity of students, which was less than 5,000 in 1960, has now gone up to 40,000. With about 350 constituent colleges, these AUs enroll, on annual basis, about 25,000 students at UG level, over 15,000 at Masters' level and Ph.D. programmes. In addition to this, there are many private affiliated colleges enrolling thousands of students annually. There are about 23,000 scientists for teaching, research and extension under the present ICAR-Agricultural Universities (AUs) system. Degree courses in 11 UG disciplines and 93 disciplines at PG level are offered with an emphasis on learning through hands on practice sessions. About 52% students admitted are from rural background and 36% are girls.

Emerging challenges in agriculture sector

Presently, agriculture faces many challenges such as (i) low productivity (averaging to 60% of world average), (ii) decreasing profitability in farming, (iii) rising quality competitiveness under the pressure of globalisation, (iv) poor linkage of farms with the market, (v) low knowledge of input agriculture, (vi) wide gap between lab and land experiments, (vii) low level of mechanization and value addition (viii) Supply Chain Management and Product Lifecycle Management, (xi) lack of qualified manpower to address the new and emerging challenges and deliver at grassroots level, (x) mounting threat to sustainability arising from depleting quality of natural resources, biotic and abiotic stresses and inefficient use of agro-inputs and (xi) poorly coordinated natural disaster management system.

To properly address these challenges, competent human resource in sufficiently large numbers would be required in the near future. To achieve this, a renewed thrust for higher agriculture education is necessary with enhanced financial support to the ICAR- AU system. Estimates suggest that by the year 2020, more than 16,000 scientific manpower would be required to cater to the needs of R & D in the country. At present, there is substantial gap of 50% or more between demand and supply of manpower in agriculture and allied sciences sector. The projections indicate that by 2020, the annual out turn required for Undergraduate and above would be about 54,000 as against the present annual out turn of around 40,000. This means that sincere efforts are required to attract more number of students towards Higher Agricultural Education. There is a vast scope for young graduates to undertake agriculture as their profession which is directly or indirectly contributing to the economic and social development of the country.

Recently, Indian Council of Agriculture Research (ICAR) has recognized degree of B.Sc. Agri in different State Agriculture Universities of the country as Professional Degree. Nine courses have been granted the recognition of 'Professional Degrees' namely, Agriculture, Horticulture, Agriculture Engineering, Sericulture, Forestry, Food Technology, Biotechnology, Home or Community Science, Food Nutrition and Dietetics. With these courses now being professional, specialized requirements can be imparted as Indian agriculture is increasingly in need of micro-level attention. Fifth Deans Committee report gave terms of reference, considering contemporary challenges for employability of passing out graduates and to adopt a holistic approach for quality assurance in agricultural education. The Committee has restructured course curricula and have reoriented the system to develop needed skills and entrepreneurial mindset among the graduates to take up self-employment, contribute to enhanced rural livelihood and food security, sustainability of agriculture and be propeller for agricultural transformation. The professional degree status provides lot of benefits for students such as enhanced rate of fellowship from DBT and other such agencies. More students and their parents prefer to study in agriculture stream.

The higher educational scenario in present times is influenced by forces of globalization, emergences of new areas of specialization such as intellectual property rights (IPRs) and World Trade Organization (WTO) and General Agreement on Trade in Services (GATS) related areas, cutting edge new beneficial technologies like biotechnology, nanotechnology etc. Agricultural graduates on completion of their studies must possess knowledge about all such new techniques. They must be able to handle the new challenges posed by globalized environment. The most important thing is that agricultural education should address the requirement of major stake holder i.e. Indian farmer.

Career scope of agricultural science and technology

India could be a leading global leader in agriculture and a significant a part of the Indian economy is driven by agriculture. Over 52% of the agricultural households depend upon agriculture for his or her employment. India is one amongst the biggest producers of vegetables, fruits and flowers within the world. Also, India is that the largest producer of milk, second- largest of sugar, leading producer of coconut and spices. India is one in all the most important exporters of agricultural commodities like spices, flowers, mushrooms, oilseeds, milk, rice, food grains and other vegetables. Society needs food, and hence, agriculture will always be a lucrative career option. Indian government is additionally backing up the agriculture sector very strongly which was clearly reflected from the budget 2016-17. Even the

2017-18 Union Budget is anticipated to pitch for more reforms in agriculture marketing and increased funds for insurance and irrigation schemes. Agriculture credit has been raised to a record Rs. 10 Lakh Crores.

Besides Government, private bodies and foreign investors also are showing great interest within the Indian agricultural industry. As per the official reports of the Department of business Policy and Promotion (DIPP), Indian agricultural sector has attracted Foreign Direct Investment (FDI) equity inflow of about USD 2.8 billion from April 2000 to March 2016. Therefore, future is certainly bright to pursue a career in agriculture.

Career opportunities in agriculture science & technology

Agricultural science involves research and development on production, processing, productivity of crops and end products for consumers. Major activities include improving quality and quantity of farming, improving crop yield, minimizing labour, conservation of soil and water and pest control. Agriculture sector isn't only about research and development. It provides roles for college kids with business backgrounds moreover. The standard career opportunities are there in Government departments, research and education, nationalized banks, agri-inputs industry and Government and political views Agencies. The emerging sectors include agribusiness, food processing, financial sector, retailing, rural marketing, international trade, rural credit and insurance, warehousing and commodities, NGOs and KPOs.

Almost ½ all the professionals within the agricultural sector have gotten business related roles. The key areas of MBA in Agri-Business and Agricultural Marketing, Agricultural Pricing, Agricultural Law, Agricultural Trading & Merchandising, Agricultural Economics, Agricultural Data Analysis and Farm Management. The non-scientific roles include marketing, technical sales, merchandising, economists, accountants, finance managers, commodity traders, communication & education.

Indian agricultural sector is facing some serious challenges further. The challenges mean there are quite enough opportunities. Many Agri-Tech Social Enterprises in India providing jobs like DuPont India, Rallies India Ltd, Nuziveedu Seeds Ltd, Lemken India Agro Equipment Pvt Ltd, Advanta Ltd, Monsanto India, Poabs Organic Estates, National Agro Industry, Godrej Agrovet Ltd, Rasi Seeds. Several startups and entrepreneurs are springing up within the agricultural sector. The four Major Areas of AgriTech Startup are disease and pest control, timely irrigation and soil health, aerial survey to detect anomalies, and packaging and transportation. Hence, persons from science and/or business background, want to try and do something good for the society, and have gotten the entrepreneurial creature , then agricultural sector is for them.

Advancements in agriculture during five year planning period (1951-2011)

The economic situation at the time of independence was quite unfavorable because of partition of India and Second World War. India was left with 82 percent population of undivided India, 69 percent of land under rice, 65 percent under wheat, 75 percent under cereals and 70 percent area under irrigation. 28 Cultivators were under heavy debt and holdings were quite uneconomic. Farmers were largely facing issue of exploitation. Therefore, government had a need of structural changes in agriculture sector. These changes has made through the economic planning and various judicious reforms during planning period. India has implemented 12 five year plans during last sixty five years and achieved great development in agriculture sector. The prime objectives of development plans have been improve standard of living and promote for social welfare, although priority has given to different issues to achieve goal of rapid economic growth and social justice during planning period in India. Therefore, period of economic planning can be divided into four phases.

The first phase of development covers the period of independence influenced both by the indigenous thinking of agricultural development and western thinking of industrial development and over during mid-sixties. The second phase consists of adoption of modern agro-technology after the mid-sixties and it continued till the announcement of new economic policy. During the period agriculture sector not only met the domestic demand of food but also surplus food production had seen in India. Moreover, rapid diversification of cropping pattern, accepting new technologies and commercialization had seen during the period of. Third phase has started with announcement of new economic policy rather than since 1985 and continued till end of 10th five year plan. The government has neglected agriculture sector and consequently new issues have born such as open market policy, reduction of agriculture subsidy, declining public expenditure on agriculture sector, farmers‟ suicide and many more during the economic reforms period. The fourth phase have initiated from 2002, in which government has focused on structural changes in agriculture sector. However, for simplicity study of agriculture development has made according to the five year plan.

I. First five year plan (1951-1956)

The approach of first plan towards agricultural development was to avoid farmers' exploitation, increase productivity of agriculture and promote farmers for increasing use of inputs and technology. Moreover, special focus was laid on land reforms, to overcome cooperative and marketing institutions from the institutional impediments for agriculture development. A community development programme had launched in 1952 to achieve the goals of plan.

Similarly, the programme of land reforms had started in 1948 in Madras state, although the constitutional amendment (Article 31) made in 1951 and cleared the way of completion of land reforms programme. This programme considers abolition of intermediaries, tenancy reforms including regulation of rent and ownership of tenants, ceiling on land holding, and consolidation of holdings and prevention on subdivision and fragmentation of land. This programme had continued till sixth five year plans.

Agriculture including irrigation and power was given top priority to achieve the goal of plan. The relative share of plan outlay on agriculture and community development, major and minor irrigation and power development was 15 percent, 16 percent and 13 percent respectively in first plan. Hence, national income increased by about 17.5 percent, per capita income by 10.5 percent, net output of agriculture and ancillary fields by 14.7 percent. Similarly, 3 million hectares of irrigation potential created and increased power generation from 2.3 million Kw to 3.4 million Kw during the first five year plan.

II. Second five year plan (1956-1961)

Top priority has given to the industrial sector to achieve goal of balanced development in second plan. Hence, share of plan outlay on industrial sector had 24 percent (Rs. 1080 crores) and 20 percent (Rs. 950 crores) on agriculture and irrigation. Therefore, programmes for agriculture development had not launched except Intensive Agricultural District programme (IADP). This programme had started in 1960-61 and subsequently extended to another thirteen districts to provide package of inputs like improved seeds, fertilizers and modern techniques for increasing productivity of agriculture sector. Moreover, decentralization of democratic power was made through setting up of Zilla Parishad and Movement Control Order was passed in 1960 to regulate the distribution of fertilizers. Establish socialistic pattern of society was also goal of second plan. Therefore, selected targets were increase national income by 25 percent, food grain production by 24 percent (65 million tons in 1955-56 to 80.5 million tons in 1960-61) and irrigation extension up to 21 million acres during the second five year plan.

III. Third five year plan (1961-1966)

The main goal of third five year plan was to be self reliant and self generating economy. Plan had given first priority to agriculture sector and aimed at doubling the growth rate of agricultural production, produce self-sufficient food grains to meet the need of nation and enough commercial crops to meet the needs of export and industry. Moreover, target of irrigation extension had 20 million acre during plan period. Hence, target of national income increase by 25 percent and agriculture production by 30 percent had selected during plan period. Unfortunately, an adverse impact on agricultural production during 1965- 66 and 1966-67 had marked due to unfavourable weather. Moreover, attack of Pakistan and China on our territory had an adverse impact on economic development during last years of plan. Therefore, national income has risen by only 2.8 per annum against the target of 5 percent per annum as well as, agriculture production has increased by 2.6 percent per annum during first four years of plan period and it has declined by 17 percent during last year due to drought.

In 1964-65 and subsequent years, modified version of Intensive Agricultural District Programme was extended in several parts of the country in 1964-65 by the name of Intensive Agricultural Area Programme. Moreover, Agricultural Prices Commission came into existence in January 1965 for giving assurance of stable prices to agriculture produces and farm income. APC was mandated to recommend minimum support prices (MSPs) to stimulate the cultivators for adopting modern technology, increase per hectare yield and overall grain production in line with the emerging demand pattern of nation.

IV. Fourth five year plan (1969-74)

Fourth plan could not be commenced after the expiry of third five year plan due to hostilities of 1962 (China) and 1965 (Pakistan), steep fall in agricultural production over two consecutive years (1965-66 and 1966-67), devaluation of rupee and inflationary pressure. These phenomenon necessitated readjustment in the planning and consequently formulated three annual plans. Growth with stability and progressive achievement of self reliance was prime aim of fourth plan. The plan has two main objectives for the agricultural sector. The first objective was to provide the necessary conditions for a sustained increase about 5 percent per annum over the next decade and second enables large section of population (small farmers, agricultural labourers, farmers of dry areas) for participating in development and share benefits. Small Farmers' Development Agency (SFDA), Marginal Farmers and Agricultural Labour Development Agency (MFAL), Command Area Development Project (CADP) and Drought Prone Area Programme (DPAP) had launched in order to participating small farmers, agriculture laborers and farmers of rain fed area to develop agriculture sector during this plan. Moreover, nationalization of 14 private commercial banks had one of the important tasks of government towards expansion and development of banking sector. Fertilizer subsidy was announced initially in 1973-74 on the imported fertilizers during this plan.

V. Fifth five year plan (1974-1979)

Fifth plan was formulated against the backdrop of severe inflationary pressure. The plan proposed to achieve the two main objectives i.e. removal of poverty and attainment of self reliance through promotion of higher growth rate, better distribution of income and step up domestic saving rate. Plan had targeted an annual growth rate of 5.5 percent national income for achieving above the goal while the actual achievement was 5.0 percent. A 20 point programme was introduced in 1975 to achieve development in various sectors. The priorities had given to increase agriculture production, irrigation level, soil and water conservation, research and education, cooperation, flood control, power sector and rural roads in this programme. Fertilizer subsidy was given initially (1973) on the imported fertilizers on account of their rising cost during this plan. Government had introduced Retention Price Scheme (RPS) for nitrogenous fertilizers in November 1977. Subsequently, RPS was extended to phosphate and other complex fertilizers from February 1979.

VI. Sixth five year plan (1979-84)

There were two sixth five year plans. The Government of Janata Party formulated Sixth Five Year Plan (1978-83) sought to reconcile the objectives of higher production with greater opportunity for employment. Focus of plan was to enlarge employment opportunities in agriculture and allied activities, to encourage household and small industries for producing consumer goods for mass consumption and raise income of lowest class through minimum needs programme. The new sixth plan (1980-85) was launched by the succeeding Government with the prime objective of direct attack on the problem of poverty by creating conditions of expanding economy. Plan had adopted strategy of strengthen infrastructure for agriculture and industrial sector. This plan had laid stress on dealing with inter related problems through a system approach rather than in separate compartments, greater and efficient management, intensive monitoring and active involvement of people. Plan had targeted growth rate of 5.2 percent per annum but actually achieved 5.4 percent. Plan had given priority for agriculture development for bridging the gap prevailing between actual and potential farm yields through the removal of constraints responsible for this gap.

The strategy of extend benefits of new technology to more farmers, cropping systems, regions and promote for greater farm management through concurrent attention for cash and non cash crops to achieve the goal of agriculture sector. Moreover, a strategy of promote scientific land water use patterns based on consideration of ecology, economics, energy, soil conservation, employment and land reforms also adopted in plan. The projected growth rate of plan had increased food grains output by 6.5 per cent per annum using a base level figure of 109 million tons of 1979-80. However, growth rate works out to 3.2 per cent. National Bank for Agriculture and Rural Development (NABARD) established in July, 1982 was the important development in the field of agriculture credit of this plan. NABARD had emerged as an apex national institution accredited with all matters concerning policy, planning and operation in the field of credit for agriculture and other economic activities of rural areas.

VII. Seventh five year plan (1985-1990)

Plan emphasized policies and programmers which aimed rapid growth in food grains production, extension of employment level and per hectare yield. A Jawahar Rogzar Yojana had launched in addition to already existing programme to reduce unemployment, regional imbalance and incidence of poverty during plan. Similarly, raising the productivity of rice in eastern region and in rainfed and dry land agriculture through extension of new agricultural programme had also prime aim of plan. A plan had targeted 4 percent annual growth rate of agriculture production. The Special Rice Production Programme for eastern region, National Watershed Development Programme for Rainfed Agriculture, scheme for assistance to small and marginal farmers had introduced to achieve targets of agriculture development in plan. A new seed policy was announced in 1988 for permitting import of high quality seed and planting material after verification of performance in Indian conditions. Similarly, third phase of National Seed Project supported by the World Bank was also launched in this plan. Besides, Comprehensive Crop Insurance Scheme has been in operation since *Kharif* 1985 for mitigating risk of farm sector. The plan had targeted 5 percent GDP growth, but actually GDP grew 5.6 percent and food grains production growth was 3.23 percent during the plan.

VIII. Eighth five year plan (1992-97)

The Eighth Plan was launched when both international and domestic market were at turning point. All centralized economies of world were disintegrating and economies of several regions are getting integrated under a common philosophy of growth guided by the market forces and liberal policies. This plan was based on autonomy and efficiency induced by competition. Generate surplus food grains production for export and achieve self sufficiency in pulses and oilseeds was the prime aim of plan. An expected annual growth rate was 4 percent in respect of gross value of agricultural output.

The strategy of more focus on dry land farming, spreading benefits of agricultural revolution to the eastern region and improving efficiency of irrigation adopted in this plan. The Eighth Plan had aimed at consolidating gains from base built over the years in agricultural production, sustaining improvements in productivity and production for meeting increasing demands of growing population; enlarging income of farmers and realizing potential of nation by stepping up agricultural export. Therefore, government had continued programmes introduced during sixth and seventh plan. The prices of phosphate and potassic fertilizers were decontrolled from 25th October 1992 during this plan. The Plan had 5.6 percent annual growth rate but actually achieved 6.7 percent, which was higher than the target.

IX. Ninth five year plan (1997-2002)

The plan has given top priority to generate adequate productive employment, eradication of poverty, price stability, assurance of food and nutritional security to all, safe drinking water, primary health care services, universal primary education, shelter, road connectivity to all, assurance of sustainable environment, empowerment of women and socially disadvantaged groups, promote public bodies for agriculture and rural development. The plan has targeted doubling the food production and make India hunger free within 10 years period. Similarly, plan focused on rising of farm productivity and other economic activities to achieve the goal of employment generation and income of rural people of India. The expected growth rate of agriculture output had 3.82 percent and agriculture sector 4.5 percent per annum. Similarly, targeted annual growth rate of livestock had 6.6 percent, poultry 7.5 percent and fisheries 6.5 percent. However, actual growth rate of GDP was 5.35 percent and 2.06 percent of agriculture sector.

X. Tenth five year plan (2002-2007)

The target of 8 percent annual GDP growth either through direct linkages that exist or through resources that were generated by growth process for more intensive public interventions had fixed for the plan. Moreover, the plan had adopted the redefining role of government in context of emergence of strong and vibrant private sector, provision of infrastructure, flexibility in fiscal and monetary policies. But actual economic growth rate was 7.77 percent per annum. The plan targeted 4 percent agricultural growth rate based on efficient use of soil and water conservation and expected benefits of agricultural development must reach to farmers of all region. Moreover, agricultural sector should have taken benefits of exports by facing challenges of economic liberalisation and globalisation. Plan had introduced new Pricing Scheme instead of Retention Price Cum Subsidy (RPS) on 1 April 2003 and adopted Tariff Committee recommendations for the complex fertilizers in this plan.

XI. Eleventh five year plan (2007-2012)

The central visions of Plan was to build up economy on their strengths and trigger a development process for ensuring broad-based improvement in quality life of SC, ST, OBC, minorities and women. A target of 9 percent growth selected that yields broad-based benefits and ensures equal opportunity to all. However, actual GDP growth rate might 8 percent per annum during plan. The plan has given priority to rainfed agriculture for increasing per hectare yield and reduces the gap of productivity of rainfed farming and irrigated farming to achieve the 4 percent selected target of agriculture growth during this plan. Similarly, Rashtriya Krishi Vikas Yojana (RKVY) had been launched during the plan period to incentivize state governments for preparing district level agricultural plans. The preparation of district-specific agricultural plan is an important innovation and expected to focus on local crop productivity constraints. Adequate funding provided to district planning under the scheme and central government had given more support to state extension systems in plan. Plan has suggested strengthen and restructure the ICAR and focus of same must be on farm oriented research. The plan focuses on more investment in agriculture rather than extension of irrigation and efficient use of water. Therefore, the target of Bharat Nirman was extended to 2.5 million hectares irrigation extension per annum by 2008-09 was fixed in this plan. In reality, average GDP growth rate of agriculture sector has achieved 3.7 percent in the plan. Status of agriculture during planning period is achieved significant development in agriculture sector in terms of production and productivity of major crops and diversification of cropping pattern. In absolute terms, 328.73 million hectare is geographical area of nation, but reporting area is 305.61 million hectare. The proportion of net sown area to reported land area has increased from 41.77 percent to 46.91 percent during the 1950-51 to 1990-91 and later declined to 45.81 percent in 2009-10. Similarly, gross cropped area has increased continuously 131.89 million hectares to 192.20 million hectares between the periods of 1950-51 to 2009-10.

Conclusion

The phase-wise development of agriculture covers the period of independence influenced both by the indigenous thinking of agricultural development and western thinking of industrial development and over during mid-sixties. It consisted of adoption of modern agro-technology after the mid-sixties and it continued till the announcement of new economic policy. In these periods agriculture sector not only met the domestic demand of food but also surplus food production in India. Moreover, rapid diversification of cropping pattern, accepting new technologies and commercialization had seen during the period of. These developments started with announcement of new economic policy rather than since 1985 and continued till end of 10th five year plan in higher rate of adaption and implementation The government has neglected agriculture sector and consequently new issues have born such as open market policy, reduction of agriculture subsidy, declining public expenditure on agriculture sector, farmers' suicide and many more during the economic reforms period afterwards.

Some pertinent short questions

1. What is the role of society in agricultural development?
2. State five groups in modern concept of agriculture in North Eastern States;
3. State career opportunities in agricultural Science & Technology;
4. Discuss agricultural development in XI plan period (2007-12);
5. What are the emerging challenges towards advancement of agricultural sciences?

5

Agricultural Resources Available Importance of Agriculture and Scope of Agriculture

Introduction

The basic resources for agriculture are sunlight, soil and water, besides the seeds and animal breeds, and human Endeavour. Another important input is the agro technique. Agricultural production is adversely affected if any of these factors is limited, or disturbed. Obviously, scientific study enables technological innovations to manipulate these factors to maximize production. Increasingly nowadays, application of science and technology also involves preventing or at least minimizing wasteful use of precious resources. The basic and natural resources in agriculture has been discussed below.

1. Soil

Optimum land utilization is important if agriculture is to be productive. India has a wide variety of soils. The Indian Council of Agricultural Research (ICAR) classifies them into eight major types. Each has its own characteristics as well as deficiencies, and each is suitable for certain crops. Soils should be tested to check their physical characteristics and nutrients before a suitable crop is grown in it: only then can yield be maximized and soil not wasted.

a) Soil Erosion

Top soil is a valuable resource—it takes nature 50 years or more to build a mere centimeter of it. Yet soil erosion is a major problem worldwide and more so in India. While all soil cannot be protected by trees and grass cover, for crop cultivation is a basic human need, cropping practice can be scientifically managed to reduce waste and erosion through (a) contour bunding—making ridges out of the soil in the field; (b) gully plugging—reducing the speed of run-off water by placing materials like hedges and sand-bags in water channels; (c) planting green manure plants and trees on field bunds; (d) control shifting cultivation which if the land is not left fallow long enough, leads to soil degradation of dangerous proportions.

b) Soil Reclamation

Growth of population and the increasing demand for agricultural products necessitates reclamation of land hitherto held as unproductive. Reduced soil productivity and abandonment of productive soil are serious issues arising out of Stalinization and alkalization. Saline and alkaline soils result from centuries of neglect, and mismanagement. Estimates of salt-affected soils in India vary, but the Central Soil Salinity Research Institute (CSSRI) puts it at nearly seven million hectares scattered all over the country.

Alkali soils contain large amounts of carbonates and bicarbonates of sodium, resulting in accumulation of sodium on the surface of soil particles and high pH. The soils have poor drainage and water transmission properties. When cultivated, clods are formed, adversely affecting standing corps. These soils mainly occur in Haryana, Punjab, Uttar Pradesh and Bihar.

Scientists at the CSSRI, Karnal, have developed effective, economical and easily acceptable technology for reclamation of alkali soils. Successful reclamation is dependent upon choosing the correct practices for specific problem soils, and involves studying the quality and depth of groundwater, calcareousness of soil, and physical condition of subsoil.

The essential components of the technologies for reclamation of alkali soils include:

(i) Proper water management through provision of irrigation through tube wells, using sprinkler method, light and frequent irrigation;

(ii) Application of proper amendments such as gypsum, sulphuric acid, sulphur and iron pyrites;

(iii) Agronomic and cultural practices like land leveling and bunding for uniform water management, effective leaching, increased dose as well as appropriate application of nitrogenous fertilizers;

(iv) Choosing the proper crops and varieties during reclamation as crops differ in tolerance to soil deficiencies, salinity, alkalinity (in partially reclaimed soil, for example, special varieties of rice can be grown as kharif, and wheat as rabi crop besides sorghum and maize);

(v) Cultivating fodder grasses and tree species like kikar (an auger hole technique small holes filled with manure to which a little gypsum is added for planting trees is effective);

(vi) Leaching along with application of gypsum; and

(vii) Drainage or removal of excess water this is accomplished when groundwater is used for irrigation, lowering the groundwater table below the critical limits of 2-3 meters.

These technologies in a suitable combination have proved effective in reclaiming alkali soils. Punjab, Haryana and Uttar Pradesh have set up Land Reclamation and Development Corporations to help farmers reclaim alkali lands lying barren for years. Significantly, it has been found that normal production levels can be attained within three to four years of reclamation, and costs recovered. Reclamation of alkali soils has environmental consequences: retention and profitable use of runoff water, and mitigation of hazards of floods.

Saline soils contain excess of neutral soluble salts including the chlorides and sulphates of sodium, calcium, and magnesium. These soils are widespread in the arid and semi-arid regions of Gujarat, Rajasthan, Haryana, Maharashtra, Kamataka, and Tamil Nadu. Salinity problems seriously affect many coastal regions due to sea water ingress and shallow saline water tables. Saline soils have a high water table, and the crucial factor in the management of such soils is drainage. The water table may be brought down below the critical limits of 2-3 meters with drainage channels one meter deep and 30 meters apart. However, success in reclamation of salt- affected soils has been confined to areas which have groundwater of relatively good quality.

In areas where groundwater has high salt content, problems of drainage continue to hamper long-term productivity of canal- irrigated areas. In this context, drainage measures must be integral to any irrigation project. Unfortunately, in the hurry to spread irrigation to a large area, this vital aspect is ignored. Coping with salinity will depend on an in-depth knowledge of soil-water-plant relationship in a hydrologic unit and the way in which the system responds to interventions for greater productivity.

2. Water

India, situated in the monsoon belt of south-east Asia, is even now dependent to a large extent, on rains for its agriculture. Rainfall is unevenly distributed, and in some years the monsoon fails, resulting in drought. It is a measure of the resilience injected into our agriculture by S & T inputs that we are not so severely affected by droughts now as in the past.

The geographic situation of the country forces us to learn to cope with inconsistent monsoons. The average annual rainfall we receive is about 370 million hectare-meters of which about 80 million hectare-meters of water seeps into the soil.

Half of this remains in the top soil and helps plant growth; the rest goes further down to constitute groundwater. However, it must be noted that saline groundwater is not useful for irrigation, so useful groundwater, according to recent reports, has depleted alarmingly.

Water is essential commodity for crop production, its deficiency acts as a bottleneck in the use of other inputs. In India, several irrigation schemes major; medium and minor have been implemented. However, the efficiency of most irrigation projects is low, causing a sizeable gap between the created and the utilized potential. Theoretically, irrigation should make possible double and multiple cropping but this is still not the case in much of the irrigated area in India. Water scarcity is an increasingly serious problem.

Obviously the farmers, by and large, lack knowledge of appropriate agronomic practices; supporting facilities necessary for optimum use of irrigation works are not maintained properly; tanks and open wells are sadly neglected. Faulty irrigation practices and absence of proper and adequate drainage facilities have been responsible not only for waste of water but for long-term damage to land through water logging, salinity and alkalinity, as well as the spread of water-borne diseases.

Irrigation Projects

There was a time when the multipurpose river valley projects were hailed because the harbingers of prosperity. Not anymore. These "temples of recent India" as national leader termed the dams, haven't any doubt proved a very important think about achieving a successful breakthrough in agriculture and a measure of self-sufficiency in food production.

However, their success has been overrated. Serious problems have arisen. Heavy siltation of the reservoirs of major dams reduces storage capacity, and that they become incapable of absorbing heavy floods. Disastrous fallout is that the degradation of the soil within the command areas of irrigation projects because of water-logging and soil salinity.

Canal irrigation, again, has contributed to land degradation. Groundwater table rises because of accession of applied water to groundwater through deep percolation form unlined field channels, distributaries and main canals, and restrictive drainage because of construction activities.

When the bottom water table rises to within 1 to 2 meters of the ground surface, it contributes significantly to evaporation form the soil surface, leaving the dissolved salts within the root zone of soils. The salt content of most groundwater within the arid and semi-arid regions is high. This further accentuates the processes resulting in the formation of salt-affected soils. There are numerous samples of land degradation following the introduction of canal irrigation.

The water from the Bhakra canal system was made available in Hissar district in Haryana in 1963. Large areas irrigated by the carnal are already salinized. Similar experiences are available within the area of the Tungabhadra irrigation project in Kamataka and within the Nagaijuna Sagar command area. The issues within the Rajasthan Canal command area have acquired serious proportions.

a) Better use of water

The emphasis on minor irrigation works in recent years is worth noting. Groundwater development is a component of this. Minor irrigation schemes, besides being a rapid and reliable source of water, also help to regulate water logging and Stalinization in canal command. States have preoccupied 'on-farm development' works altogether the command areas and are introducing intermittent water delivery methods to cut back water application even while increasing farm production.

Sprinkler irrigation is water-efficient, and has been introduced in canal irrigated areas of Haryana, Rajasthan and Madhya Pradesh. Drip irrigation is being encouraged in Maharashtra, Kamataka, Andhra Pradesh, Orissa, and province for fruit crops. 'Fertigation' is an innovation: the appliance of fertilizer through drip irrigation to maximize farm productivity with available water.

As for surface irrigation, scientists have established that paddy fields move with a way lower depth of standing water than commonly in practice. Methods to conserve water include intermittent water delivery, furrow method, micro-irrigation in well- irrigated areas, and delivery of water in pipelines to avoid loss on way. The green-house method is another scientific way of growing special crops.

3. Seeds

There was a time once you could get certain vegetables in certain seasons only. Not now: nowadays you get cabbages and cauliflowers practically the entire year through. By genetic manipulations, scientists have led to change within the genetic architecture of plants and within the development rhythms of plants so they're not "season bound" but "period fixed". These new varieties are fitted in rotation with other crops, or the simultaneous cultivation of crops can increase production and productivity.

Scientists create ideotypes, or plants with ideal frame and desirable characteristics, chosen after scanning the wide variations within the available germplasm. By incorporating several of the desirable characters in one variety they create a hybrid which can be not only high-yielding but also pest- and disease-resistant, and of short duration.

There are three stages within the seed production cycle for efficient implementation of the high yielding varieties (HYV) programme. Breeder seed is that the primary stage. it's multiplied into foundation seeds to be multiplied finally into certified (quality) seeds which are distributed to any or all farmers all told regions.

The government founded the National Seeds Corporation (NSC) in 1963 and also the State Farms Corporation of India (SFCI) and State Seed Corporations (SSCs). They produce and provide quality seeds. The ICAR organizes production of breeder seeds with the breeders' cooperation with the scientists concerned. Seed technology research was initiated in an exceedingly coordinated manner in 1979 on receipt of indents from state governments.

The National Seeds Project clinical test (NSP-III), launched in March 1990 with World Bank assistance, led to June 1996, but its residual activities are still continuing. NSP-III had two components (a) project component under which funds were released for (i) investment credit to seed sector through NABARD; (ii) varietal development and breeder seed production through ICAR; (iii) strengthening of seed certification and testing agencies; (iv) institutional strengthening; and (b) programme component under which funds were released to State Seed Corporations, NSC and SFCI for his or her financial and organizational restructuring.

The scheme entitled National Programme for Varietal Development was initiated in 1995-96 during the Eighth Five Year Plan and was continued up to 1997-98 during the Ninth Five Year Plan.

However, from 1999-2000, the scheme was revised and a replacement scheme entitled Establishment and Maintenance of Seed Bank was launched with the essential objectives of meeting requirement of seed during natural calamity and also development of necessary infrastructure for storage of seed. The scheme is being implemented through the NSC, SFCI and SCIs of several states.

Development of status

a) Potential resources

Resources are found in a very region but haven't been utilized. as an example, the western parts of India particularly Rajasthan and Gujarat have enormous potential for the event of wind and solar power, but to this point these haven't been developed properly.

b) Developed resources

Resources which are surveyed on their quality and quantity are determined for utilization. The event of resources depends on technology and level of their feasibility.

c) Stock

Materials within the environment which have the potential to satisfy human needs but groups of people don't have the suitable technology to access these, are included among stock. for instance, water may be a compound of two inflammable gases; hydrogen and oxygen, which might be used as an upscale source of energy. The specified technical 'know-how' to use them for this purpose. Hence, it is considered as stock. Reserves are the subset of the stock, which might be put into use with the assistance of existing technical 'know-how' but their use has not been started. These may be used for meeting future requirements. River water will be used for generating hydroelectric power but presently, it's being utilized only to a limited extent. Thus, the water within the dams, forests etc. may be a reserve which may be employed in the long run.

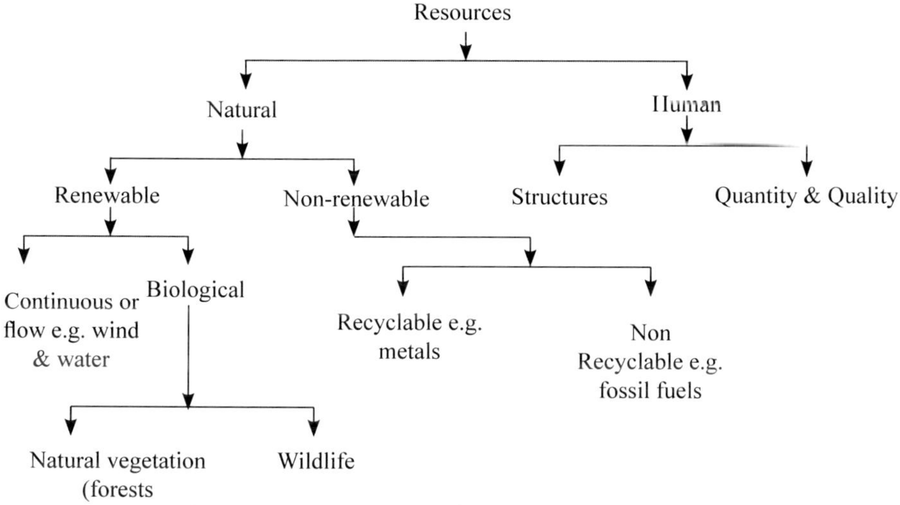

Fig 1: Classification of resources

Idea of origin

a) Biotic Resources: These are obtained from biosphere and have life like mortals, flora and fauna, fisheries, livestock etc.

b) Abiotic Resources: All those things which are composed of non living things are called abiotic resources for instance, rocks and metals.

c) On the idea of Exhaustibility

d) Renewable resources

e) The resources which may be renewed or reproduced by physical, chemical or mechanical processes are called renewable resources as an example, solar and wind energy, water, forests and wildlife, etc. The natural resources may further be divided into continuous or flow (Fig.1). Non-Renewable Resources: These occur over an awfully long geologic time. Minerals and fossil fuels are samples of such resources. These resources take immeasurable years in their formation. A number of the resources like metals are recyclable and a few like fossil fuels can't be recycled and find exhausted with their use.

Premise of ownership

Individual resources

These are owned privately by individuals. Many farmers own land which is allotted to them by government against the payment of revenue. In villages there are people with land ownership but there are many that are landless. Urban people own plots, houses and other property. Plantation, pasture lands, ponds, water in wells etc. are a number of the samples of resources ownership by individuals. Make an inventory of resources owned by your household.

Community owned resources

There are resources which are accessible to all or any the members of the community. Village commons (grazing grounds, burial grounds, village ponds, etc.) public parks, picnic spots playgrounds in urban areas are actual accessible to all or any the people living there.

National resources

Technically, all the resources belong to the state. The country has legal powers to amass even holding for public good those are seen roads, canals, railways being constructed on fields owned by some individuals.

Urban Development Authorities get empowered by the govt. to accumulate land. All the minerals, water resources, forests, wildlife, land within the political boundaries and oceanic area up to 12 nautical miles (22.2 km) from the coast termed as territorial water and resources therein belong to the state.

International resources

There are international institutions which regulate some resources. The oceanic resources beyond 200 nautical miles of the Exclusive Economic Zone belong to open ocean and no individual country can utilize these without the concurrence of international institutions.

Land resources

We survive land, we perform our economic activities ashore and that we use it in several ways. Thus, land may be natural resources of utmost importance. It supports natural vegetation, wild life, human life, economic activities, transported communication systems. However, land is an asset of a finite magnitude, therefore, it's important to use the available land for various purposes with careful planning.

India has land under a spread of relief features, namely; mountains, plateaus, plains and islands. About 43 per cent of the expanse is apparent, which provides facilities for agriculture and industry. Mountains account for 30 per cent of the full expanse of the country and ensure perennial flow of some rivers, provide facilities for tourism and ecological aspects.

About 27 per cent of the world of the country is that the plateau region. It possesses rich reserves of minerals, fossil fuels and forests

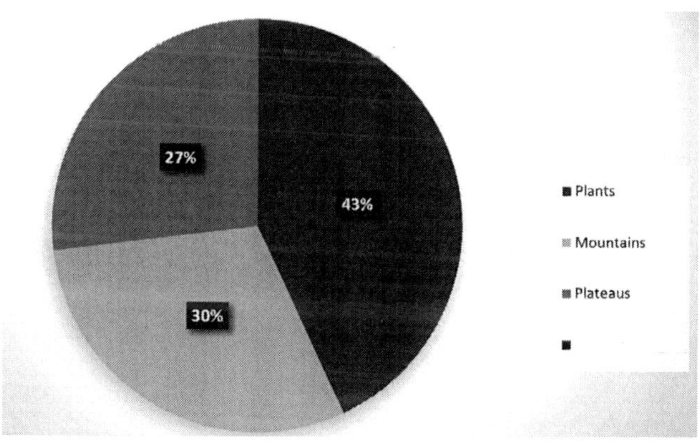

Fig 2: Land Resources

Land utilization

Land resources are used for the subsequent purposes:

1. Forests

2. Land not available for cultivation

(a) Barren and waste land

(b) Land put to non-agricultural uses, e.g. buildings, roads, factories, etc.

3. Other uncultivated land (excluding fallow land)

(a) Permanent pastures and pastureland,

(b) Land under miscellaneous tree crops groves (not included in net sown area)

(c) Cultivable waste land (left uncultivated for quite five agricultural years).

4. Fallow lands

(a) Current fallow-(left without cultivation for one or but one agricultural year),

(b) apart from current fallow- (left uncultivated for the past 1 to five agricultural years).

5. Net sown area

Area sown quite once in an agricultural year plus net sown area is understood as gross cropped area.

Land use pattern

The use of land is decided both by physical factors like topography, climate, and soil types moreover as human factors like population density, technological capability and culture and traditions etc.

Total geographic area of India is 3.28 million sq km. Land use data, however, is offered just for 93 per cent of the full region because the land use reporting for many of the north-east states except Assam has not been done fully. Moreover, some areas of Jammu and Kashmir occupied by Pakistan and China have also not been surveyed. The land under permanent pasture has also decreased. How are we ready to feed our huge cattle population on this pasture land and what are the implications of it? Most of the opposite than this fallow lands is either of poor quality or the value of cultivation of such land is extremely high. Hence, these lands are cultivated once or twice in about two to a few years and if these are included within the net sown area then the share of NSA in India involves about 54 per cent of the entire reporting area.

The pattern of net sown area varies greatly from one state to a different. It's over 80 per cent of the whole area in Punjab and Haryana and fewer than 10 per cent in Arunachal Pradesh, Mizoram, Manipur and Andaman Nicobar Islands.

Forest area within the country is much not up to the specified 33 per cent of geographic region, because it was outlined within the National Forest Policy (1952). It had been considered essential for maintenance of the ecological balance. The livelihood of immeasurable folks that carry on the fringes of those forests depends upon it. Part of the land is termed as waste land and land put to other non-agricultural uses.

Waste land includes rocky, arid and desert areas and land put to other non-agricultural uses includes settlements, roads, railways, industry etc. Continuous use of land over a protracted period of your time without taking appropriate measures to conserve and manage it's resulted in land degradation. This, in turn, has serious repercussions on society and therefore the environment.

Soil as a resource

Soil is that the most vital renewable resource. It the medium of plant growth and supports differing kinds of living organisms on the world. The soil could be a living system. It takes legion years to create soil up to some cm exhaustive. The bed rock, climate, vegetation and other varieties of life and time are important factors within the formation of soil.

Various forces of nature like change in temperature, actions of running water, wind and glaciers, activities of decomposers etc. contribute to the formation of soil. Chemical and organic changes which happen within the soil are equally important. Soil also consists of organic (humus) and inorganic materials. On the idea of the factors to blame for soil formation, colour, thickness, texture, age, chemical and physical properties, the soils of are often classified in several types.

Renewable and non-renewable resources

Natural resources and has associated problems with non-renewable resources and renewable resources. These are discussed as under-

a. **Forest resources:** Use and over-exploitation, deforestation, case studies. Timber extraction, mining, dams and their effects on forests and tribal people

b. **Water resources:** Use and over-utilization of surface and spring water, floods, drought, conflicts over water, dam benefits and problems.

c. **Mineral resources:** Use and exploitation, environmental effects of extracting and using natural resources, case studies.

d. **Food resources:** World food problems, Changes in land use by agriculture and grazing, Effects of contemporary agriculture, Fertilizer/ pesticide problems, Water logging and salinity

e. **Energy resources:** Increasing energy needs, Renewable/ nonrenewable, Use of Alternate energy sources, Case studies

f. **Land resources:** Land as a resource, land degradation, man-induced land-slides, wearing and desertification. Ecosystems act as resource producers and processors. Solar power is that the main actuation of ecological systems, providing energy for the expansion of plants in forests, grasslands and aquatic ecosystems. A forest recycles its stuff slowly by continuously returning its dead material, leaves, branches, etc. to the soil. Grasslands recycle material much faster than forests because the grass dries up after the rains are over each year. All the aquatic ecosystems are alternative energy dependent and have cycles of growth when flowers spread and aquatic animals breed. The sun also drives the water cycle. Our food comes from both natural and agricultural ecosystems. Traditional agricultural ecosystems that relied on rainfall are modified in recent times to provide more and more food by the addition of additional chemicals and water from irrigation systems but are still captivated with solar power for the expansion of crops. Moreover modern agriculture creates a range of environmental problems, which ultimately result in the formation of unproductive land. These include irrigation, which ends up in the event of saline soil, and therefore the use of artificial fertilizers eventually ruin soil quality, and pesticides, which are a peril for humans moreover as destroying components vital to the long-term health of agricultural ecosystems. To manufacture consumer products, industry requires raw materials from nature, including water, minerals and power. During the manufacturing process, the gases, chemicals and waste products pollute our surroundings, unless the industry is carefully managed to scrub up this mess.

g) **Water resources**: The water cycle, through evaporation and precipitation, maintains hydrological systems which form rivers and lakes and support in an exceedingly kind of aquatic ecosystems. Wetlands are intermediate forms between terrestrial and aquatic ecosystems and contain species of plants and animals that are highly moisture dependent. All aquatic ecosystems are utilized by an outsized number of individuals for his or her daily needs like drinkable, washing, cooking, watering animals, and irrigating fields. The globe depends on a limited quantity of water. Water covers 70% of the earth's surface but only 3% of this are often H2O. Of this, 2% is in polar ice caps and only one is usable water in rivers, lakes and subsoil aquifers. Only a fraction of this may be actually used. At a worldwide level 70% of water is employed for agriculture about 25% for industry and only 5% for domestic use. However this varies in numerous countries and industrialized countries use a greater percentage for industry. India uses 90% for agriculture, 7% for industry and three for domestic use. One in every of the best challenges facing the globe during this century is that they have to rethink the management of water resources. The planet population has passed the 6 billion mark. Supported the proportion of adolescents in developing countries, this may still increase significantly during the subsequent few decades. This places enormous demands on the world's limited freshwater supply. The overall annual freshwater withdrawals today are estimated at 3800 cubic kilometers, twice the maximum amount as just 50 years ago. Studies indicate that an individual needs a minimum of 20 to 40 liters of water per day for drinking and sanitation. Over one billion people worldwide haven't any access to scrub water, and to several more, supplies are unreliable. Local conflicts are already spreading to states. eg:. Karnataka and province over the waters of the Krishna. India is anticipated to face critical levels of water stress by 2025. At the world level 31 countries are already in need of water and by 2025 there'll be 48 countries facing serious water shortages. The UN has estimated that by the year 2050, 4 billion people are going to be seriously laid low with water shortages. This may result in multiple conflicts between countries over the sharing of water. Around 20 major cities in India face chronic or interrupted water shortages. There are 100 countries that share the waters of 13 large rivers and lakes. The upstream countries could starve the downstream nation's resulting in political unstable areas across the planet. Examples are Ethopia, which is upstream on the Nile and Egypt, which is downstream and highly obsessed on the Nile. International accords which will examine a good distribution of water in such areas will become critical to world peace. India and Bangladesh have already got a negotiated agreement on the water use of the Ganges.

h) **Irrigation resource:** Apart from the final way of flood-flushing practiced at different sites of Indus Civilization, mention could also be made from the subsequent practices particularly undertaken in Sind. This is often evident within the writings of inscriptions on seals from Mohenjo-Daro which state: 'Two canals were constructed within the month of Kudam (Aquarius) and Mina (pisces). Another seal records that it took 'thirteen months to construct a canal'. These canals were cut at low gradient and waters were allowed to merge ultimately into low ground in a very direction oblique to the river so on secure a good fall as possible. This fall varied from a foot to some inches during a mile. They were up to hundred feet wide and a couple of dozen feet deep, not enough to draw removed from the river except overflowing. These were natural streams and possibly seasonal. The perennial canals of the Indus system consistent with R.B. Buckley, were move an additional depth and will receive water even when the most rivers was low, but their real purpose seems to extend the discharge during inundation. Irrigation system was consisting of lifting of water by water-wheel and by means of Shade from adjoining water sources. There was also provision of tank irrigation. Rice cultivation shows necessary moisture of soil probably retained by now using contour bunding for locking up monsoon rain-waters by tank-irrigation.

Conclusion

The basic resources for agriculture are sunlight, soil and water, besides the seeds and animal breeds, and human Endeavour. Most important input is the adoption of agro technique to enhance crop production and agricultural productivity. Agricultural production is adversely affected if any of these factors is limited, or disturbed or discontinued from the sources. Hence, scientific study for resource management enables technological innovations to manipulate these factors to maximize production. Increasingly nowadays, application of science and technology also involved preventing or at least minimizing wasteful use of precious resources.

Some pertinent short questions

1. Name five basic natural resources available for agriculture;
2. Differentiate renewable and non-renewable natural resources;
3. State present land use pattern in our country;
4. Illustrate land resources of our country;
5. What do mean by judicious use of water resources?

6

Importance of Farmers and Ancient Agricultural Practices and Scientific Basis, Status of Agriculture as Advice by Sages to Kings on Their Duties Towards Farmers

Introduction

Evolution of agriculture in India has an ancient tradition of farming since beginning of human civilization. It has been referred in several ancient scripts including Vedic literature and many other ancient scriptures. Therefore, the study consists methods of cultivation, cropping pattern and status of agriculture sector in economy and in human life is essential for strengthen the agriculture sector during recent period. The agricultural sector in pre-independence India was very prone to breakdown and highly unstable. In other words, there was not stable production rate for various crops. The main reason for this was improper infrastructure for perennial irrigation, non-judicious water use, imbalance use fertilizers and manures, uncertainty and uneven distribution of rainfall throughout the year.

Importance of the farmers

Farmers, who put their labour in crop cultivation, grow crops in the land, supply food to the people generation to generation, most unprivileged and unorganized sector of their communities who used to sale labour. In ancient period, workforces were the actual owner of the land and their produce. Gradually, the farmers became paid workforce in the production system.

In the Sanskrit literature described "agriculture is a divine skill taught by sages to farmers for sustenance of life" and shows the high regard for farmers in society. The farmer was the *Annadata* that stands for the food provider. Farmers raised fodder for animals. Farmer planted and maintained the orchard of blessed trees after that to the village, maintained pasture land, grew herbs and medicinal plants for supply to the village physician, and fragrant flowers for worship at the temple. Farmers nurtured cattle, which gave milk and carried burdens for the community. Farmers, no doubt were the source of raw materials for all artisans and ever increasing population of the country. But farmers are often neglected, exploited, cheated and deprived from their own legitimate cash out. Farmers' rights were slowly whittled down at first by foreign invasions, later by changes in the land ownership.

Farmer in village administration

Up to the middle of the sixteenth century, farmers continued to have complete rights over their lands. The rulers had claimed to a portion of the grain produced, but had no title to the farmers land. The entire administration of the village community was in the hands of groups of farmers called' *Kutumbin'* in Sanskrit, and by different names in different regions. Ancient texts describe the *Gram Sabha* in every village in which farmers elected office-bearers to look after the administration and welfare of the community. Thus farmers were responsible for their own administration. The *Gram Sabha* was noted for its prompt action to settle disputes and administer justice. Running schools and build houses for the poor, rest houses for travellers, maintaining wells and tanks for potable water supply, building bunds for protection from floods and even defence against enemy's attack were part of its duties.

i. Agriculture during Vedic period in India (3000 BC to 200 AD)

Vedas, Upanishads and Puranas are the oldest literatures of human civilization rather than important literatures of Vedic period. The scholar suggests that dates regarding the composition of Vedas right from 10000 BC to 1300 BC. Generally, Vedic period divided into early Vedic period (3000 BC to 1300 BC) and later Vedic period (1200 BC to 26 BC). The former is known as Bronze Age and later is an Iron Age. Vedas, Upanishads, Puranas and books written by Maharishi Jaimini, Rishi Parashara, Maharishi Kashyap, Rishi Varahmihir, Surupala and many others has footprints of farm business. Rig Vedic economy was sustained by a combination of pastoralism and agriculture. Cow was the centre of the Vedic village economy and in the centre of the Krisi. Rig Veda has a reference of field levelling, seed processing and storage of grains in large jars. Almost 300 varieties of various trees have been listed in *Atharvaved*. The Vedic society had transformed from semi nomadic life to settled agriculture in the later Vedic period. It leads to an increase in trade and competition for resources in Indian subcontinent. The base of krisi had accept and kept trust on rhythm of nature. Agriculture was mostly dependent on rainfall and wood plough was the prime instrument of cultivation during this period. Agriculture was dominant economic activity along the Ganges valley during this period. Paddy, barley and wheat were the main crops during later Vedic period. The Serial "Mahabharata" represents about major crops of Vedic period. The famous story "Pohe of Sudama" (**Sudama** is also known as **Kuchela** was a childhood friend of Hindu deity Krishna from Mathura) indicates that paddy was the main food grain crop and processing on paddy also known to him during Vedic period. Moreover, *loni* and *roti* had used as *shidori* while going on the field for cattle raring indicates barely or jowar was major crops during same period. Rishi Parashara (400 BC) has written a book "Vrukshaurved." This book deals with the measurement of rain, its length, rain forecasting and field crops. Moreover, various types of tree plantation, rising of nursery, grafting, transplanting, medicinal use of different plants and its application as pesticides are also described in this book. In the words of Parashara, "farmers yield gold if properly managed farm but lead to poverty if neglected farming." He has pointed out continuous assessment of cultivation methods and research on farming is required for agriculture development. The book written by Sage Kashyapa (800 AD) namely "Krishisukti" deals with agriculture planning and characteristics of good farmers. He wrote in his book, kings must build deep reservoir of water and it should be regularly examined by him. Planting of trees is important, because it gives pleasure to birds and fruits, flowers and wood to human. Moreover, farmers should be virtuous, spirituals, free from shadripus and hospitable to guests. Farmers should be careful while utilizing natural resources. Farmers must be free from hypocrisy and jealousy. They must be efficient in calculation and sound in behaviour. Vedic and ancient Indian civilization was developing along river valleys. This civilization was well developed irrigation system. It was playing crucial role in production of food grains and cotton. Vedas refers Avata (water wells), Kulya (canal) and Sarsi (dam). It indicates Vedic people were known about the devices of irrigation. Manu mentions the takata or artificial storage of water. Mahabharata contains indication of irrigation from wells, tanks and canals. Kautilya says, "If privately managed dams are neglected for 5 years, their charge is taken over by the state. If water revisers constructed by public state, revenue is to be remitted for 5 years. If only repairs are carried out by public efforts revenue is to be remitted for 4 years. Magasthenis (400 BC) mentioned that the whole country was under irrigation and very prosperous because of double harvests. Surupala a physician (1000 AD), wrote "Vrikshayurveda" (Vriksha means plants and Ayurveda means science of life). He explained importance of trees in human life such as "Planting of plants is far better than giving birth to 10 Childs". Moreover, the trade with Asian countries had increased during the later Vedic period and economic prosperity has explained in the concurrent literatures. Therefore, gold, silver, precious metals had come with exchange of the agriculture product and agro processed goods in India. There are various views regarding the land tenure during Vedic and later Vedic period. Manusmiriti represents, "land is the property of him who cut away the wood or who tilled and cleared it." Yajnavalkya says that land belongs to the king and the tiller possesses only the utilitarian right of life. Arthashastra mentioned that land may be confiscated from those who do not cultivate it and given to others or it may be cultivated by village labourers and might pay less. In Jaimini"s view, "the king cannot give away the earth because it is not his exclusive property but it is common of all human beings to enjoy the fruits of their own labour in it." All these views indicate that the state ownership was not totally absent, but limited by or combined with common ownership of the village. The farming and animal husbandry had dominant economic activities during the Vedic period. Paddy, barely, sugarcane, sesame, mustard, green gram, black gram were conventional crops and cotton, jute, sandal, camphor saffron, turmeric, fenugreek, cumin, fennel, black pepper, cardamom and clove were high value cash crops during the later Vedic period. The food grain production was sufficient rather than surplus during this period. Finally, cow raring was dominant than any other business, because cow dung had been used as fertilizers and medicine during Vedic period.

ii. Agriculture during medieval period in India (200 AD to 1757 AD)

This period can be divided into two parts i.e. early medieval and later medieval. The early medieval period is indentified as prosperity of Indian economy. However, a regular war happened between the various kingdoms which created many problems before farmers and traders. Systematic ploughing, manuring, weeding, irrigation and crop protection were practiced for sustained agriculture and water storage systems were designed properly in this period. It has positive impact on the agriculture diversification towards commercialization and processing of agriculture produces. Wheat, jowar, barely, pulses were the major crops. As well as, spices particularly black pepper, tobacco, cotton, jute, sugarcane and turmeric were the exportable cash crops produced during the medieval period in India. These products mostly used for export to Arabian countries to exchange with precious metals and horses. Zamindari was developed as a state right into kings superior ownership of the entire domain, but the concurrent hereditary, permanent and long established right of the *khu-kasht raiyats* was recognized during Muslim rule. Land revenue was collected not only officials of state but by local chiefs or lieges with whom settlements were made by the central state authorities. The *Subedars, Nawabs*, and Princes were made settlements with the lower lieges. The process of land revenue was initiated under Sher Shah and perfected under Emperor Akbar. Almost 1/3 of the 10 years average produce was fixed as the state revenue payable in cash in Akbar's regime. Great attention had been given towards irrigation development during the Pre-Mughal and Mughal period. A dam built on river Kaveri is considered one of the oldest water reservoir structures of the world. The western Jamuna Canal is known as Firoze Shah Tuglak had built for brought water to Delhi during 14th century in north India. The Bari Boab canal was executed by Ali Mardan Khan during the 17th century. The Eastern Jamuna Canel was built by the Mohamed Shah during the 18th century. It means large efforts had made for irrigation development during medieval age in Indian subcontinent. The essence is that agriculture sector had also dominant place and trade with the entire Asia and Europe bring economy in the phase of economic prosperity during the early medieval period in India. But, regular wars happened between the states and losses made by the militants in north and central region had adversely affected agriculture sector during later medieval period in India.

Kings, Sages and Saints

A prosperous agriculture being the base of strong kingdoms or empires, it was almost always supported by the multitudes of Indian rulers. The tradition was to impose minimal tax on farmers, rarely exceeding one-sixth of the produce. That was being probably need to continuously reminded of the wisdom of the ancestors and provide genuine respect and importance to farmers.

The kingdom of **Magadha** became the most powerful amongst the sixteen states that dominated this transitional period, over the time. At the outset, it was just one of eleven located up and down the Ganges River. The rest were established in the older northwest or central India. In general, larger kingdoms dominated the Ganges basin while smaller clan-based states thrived on the periphery. The kings all fought with each other over land and resources, making this a time of war and shifting alliances. The victors were the states that could field the largest armies. As such the rulers had to mobilize the resources of their realms. The Magadhan kings did this most effectively. Expansion began in 545 BCE under King Bimbisara. His kingdom was small, but its location to the south of the lower reaches of the Ganges River gave it access to fertile plains, iron ore, timber, and elephants. Governing from his inland fortress at **Rajagriha**, Bimbisara built an administration to extract these resources and used them to form a powerful military. After concluding marriage alliances with states to the north and west, he attacked and defeated the kingdom of Anga to the east. His son Ajatashatru, after killing his father, broke those alliances and waged war on the Kosala Kingdom and the Vrijji Confederacy. Succeeding kings of this and two more Magadhan dynasties continued to conquer neighboring states down to 321 BCE, thus forging an empire; its reach was largely limited to the middle and lower reaches of the Ganges River.

After existences of individual ownership of land and division of labour in the society the caste system created in the interest of the rulers. As the population of northern India rose and the landscape was dotted with more villages, towns, and cities, society became more complex. The social life of a Brahmin priest who served the king differed from that of a blacksmith who belonged to a town guild, a rich businessman residing in style in a city, a wealthy property owner, or a poor agricultural labour living in a village. Thus, the social identity of each member of society differed.

In ancient India, one measure of identity and the way people imagined their social life and how they fit together with others was the *varna* system of four social classes, *Brahmin, Khatriya, Baisya* and *Shudra*. Another came as that was **caste**. Like the *varnas*, castes were hereditary social classifications; unlike them, they were far more distinct social groups. The four- fold *varna* system was more theoretical and important for establishing clearly who the powerful spiritual and political elites in society were: the *Brahmins* and *Kshatriya*. But others were more conscious of their caste. There were thousands of these, and each was defined by occupation, residence, marriage, customs, and language. The sensitive issues like hatred, apartheid and untouchable and unprivileged section of the society came to exist. Thus, that is how caste defined an individual's life. The lowest castes were the untouchables. These were peoples who engaged in occupations considered highly impure, usually because they were associated with taking life; such occupations include corpse removers, cremators, and sweepers. So those who practiced such occupations were despised and pushed to the margins of society. Because members of higher castes believed touching or seeing them was polluting, untouchables were forced to live outside villages and towns, in separate settlements. Hence, amongst the so many good these are also of the Indian heritage that influenced agriculture from ancient period.

Buddhism came to challenge **Brahmanism and** during this time of transition, some individuals became dissatisfied with life and chose to leave the everyday world behind. Much like the sages of the *Upanishads*, these renunciants, as they were known, were people who chose to renounce social life and material things in order that they might gain deeper insight into the meaning of life. Some of them altogether rejected Brahmanism and established their own belief systems. The most renowned example is **Siddhartha Gautama** (c. 563 – 480 BCE), who is otherwise known as the **Buddha**. Buddha means "Enlightened One" or "Awakened One," implying that the Buddha was at one time spiritually asleep but at some point woke up and attained insight into the truth regarding the human condition. His life story is therefore very important to Buddhists, people who follow the teachings of the Buddha. Siddhartha was born a prince, son to the chieftain of Shakya, a clan-based state located at the foothills of the Himalaya in northern India. His father wished for him to be a ruler like himself, but Siddhartha went in a different direction. At his age of twenty-nine, after marrying and having a son, he left home. Legends attribute this departure to his having encountered an old man, a sick man, a corpse, and a wandering renunciant while out on an excursion. Aging, sickness, and death posed the question of suffering for Siddhartha, leading him to pursue spiritual insight. Goutam Buddha sought instruction from other wanderers and experimented with their techniques for liberating the self from suffering through meditation and asceticism since year. But he failed to obtain the answers he sought.

Then, one day, while seated beneath a tree meditating for an extended period of time, a deepening calm descended upon Siddhartha, and he experienced *nirvana*. He also obtained insight into the reasons for human suffering and what was needful to end it. This insight was at the heart of his teachings for the remaining forty-five years of his life. He travelled around northern India teaching his dharma that was his religious ideas and practices and gained a thousand and thousand followers and disciples as following of his students.

After the Buddha died c. 480 BCE, leaving behind thousand of his disciples who established monastic communities known as the Buddhist *sangha*. Regardless of their *varna* or caste, both men and women could choose to leave home and enter a monastery as a monk or nun. They would shave their heads, wear ochre-colored robes, and vow to take refuge in the Buddha, dharma, and *sangha*. Doing so meant following the example of the Buddha and his teachings on morality and meditation, as well as living a simple life with like-minded others in pursuit of *nirvana* and an end to suffering.

The most powerful amongst this succession of states was the **Kushan Kingdom**, whose origins take far away to the north of China. There, in the second century CE, nomadic groups struggling with scarcity moved west, displacing another group and forcing them into northern Afghanistan. Those peoples are known as the Yuezhi (yew-eh-jer), and they were made up of several tribes. In the first century CE, a warrior chieftain from one Yuezhi tribe, the Kushans, united them, invaded northwest India, and assumed exalted titles befitting a king. His successor, ruling from Afghanistan, gained control over the Punjab and reached into the plains of the upper Ganges River.

The *Upanishads*, however, added an entirely new set of ideas. The title means "sitting near" and points to a setting where sages conveyed spiritual insights to students through dialogue, stories, and analogies. The *Upanishads* are records of what was taught and discussed, the earliest dating to the eighth and seventh centuries BCE. These sages were likely hermits and wanderers who felt spiritually dissatisfied with the mythological and ritualistic approach of Brahmanism. Rather, they sought deeper insight into the nature of reality, the origins of the universe, and the human condition. The concepts that appear throughout these records of the outcome of their search are ***brahman*** (not to be confused with Brahmins), ***atman***, **transmigration**, and **karma**.

According to these sages, human beings face a predicament. The universe they live in is created and destroyed repeatedly over the course of immense cycles of time, and humans wander through it in an endless succession of deaths and rebirths. This wandering is known as transmigration, a process that isn't random, but rather determined by the law of karma. According to this law, good acts bring a better rebirth, and bad acts a worse one. It may not happen in this lifetime, but one day virtue will be rewarded and evil punished.

Ultimately, however, the goal is to be liberated from the cycle of death and rebirth. According to Hindu traditions, the *Upanishads* reflect spiritual knowledge that was revealed to sages who undertook an inward journey through withdrawal from the world and meditation. They once discovered that one divine reality underlies the universe. They called this ultimate reality *brahman*. They also discovered that deep within the heart of each person lies the eternal soul. They called this soul *atman*. Through quieting the mind and inquiry, the individual can discover *atman* and its identity with *brahman*: the soul is the divine reality. That is how a person is liberated from the illusion of endless wandering.

In conclusion, by the end of the Vedic Age, northern India had undergone immense changes. An Aryan civilization emerged and spread across the Indo-Gangetic Plains. This civilization was characterized by the Brahmin's religion (Brahmanism), the use of Sanskrit, and the *varna* social system. The simpler rural life of the clans of earlier times was giving way to the formation of states, and new religious ideas were being added to the evolving tradition known today as Hinduism. On the pretext of social changes took place, the agricultural heritage continue to generation to generation. Agriculture, always be the source of maintain livelihood. The Emperors (Kings), always be the chief of the territory where sages and saints were only link to the kings to give direction on development of agriculture and travellers (*parjatak/paribrajak*) from different countries use to bring seed, seed materials from one location to other.

Such one, greatest Kushan ruler, **King Kanishka**, furthered what these first two kings began, forging an empire extending from Central Asia across the mountain ranges of Afghanistan into much of northern India. Ruling the many peoples of such a sprawling territory required more than the periodic plundering campaigns of nomad chieftains. One sculpture of King Kanishka puts these Central Asian roots on display. In it, he is wearing a belted tunic, coat, and felt boots, and carrying a sword and mace. Kushan gold coins, however, cast him and his two predecessors in another light: as universal monarchs. On one side, the crowned kings are displayed along with inscriptions bearing titles used by the most powerful Indian, Persian, Chinese, and Roman emperors of that time. The obverse side contains images of both Indian and foreign deities.

The Kushan rulers, it appears, solved the problem of ruling an extensive, culturally diverse realm by patronizing the many different gods beloved by the peoples living within it. Buddhists, for instance, saw King Kanishka as a great Buddhist ruler, much like they did King Ashoka. In fact, Kanishka supported Buddhist scholarship and encouraged missionaries to take this faith from India to Central Asia and China. But his coins also depict Greek, Persian, and Hindu deities, suggesting that he was open-minded, and perhaps strategic, in matters of religion.

After Kanishka's reign, from the mid-second century CE onwards, the empire declined. Like the other, larger Indian states during this time, only a core area was ruled directly by the king's servants. The other areas were governed indirectly by establishing tributary relations with local rulers. As Kushan power waned, numerous smaller polities emerged, turning northern and central India into a mosaic of states.

The Indian peninsula, the territory south of the Indo-Gangetic Plain and the Vindhya Mountain Rang was also the features more prominently after the fall of the Mauryan Empire. In the south, kingdoms emerged for the first time. The largest was the **Satavahana Kingdom**, which included most of the Deccan Plateau and lasted about three centuries. The first rulers were former Mauryan officials who capitalized on its dissolution, established their own state, and expanded to the north. To establish their legitimacy, Satavahana kings embraced Aryan civilization by allowing Brahmins to perform sacrifices at the court and by upholding the *varna* social order. They also prospered from a rich agricultural base and trade. However, like so many of the larger states during these centuries, this kingdom was only loosely integrated, consisting of small provinces governed by civil and military officers and allied, subordinate chieftains and kings.

iii. Agriculture during colonial British period (1757 to 1947)

The footprints of agriculture of British period have been seen in the literature written by Indian freedom fighters, socialists, tourists and reports of British government. The Indian market is extended throughout entire world. Invention of new sea route and industrial revolution had a positive and negative impact on Indian economy. The British East India Company had played a role of traders between the periods of 1600 AD to 1757 AD in India. But, company entered as a ruler from 1757 in Bengal state and till 1818 brought entire India under the rule of company. The company had interested in commerce and collection of ransom from various Kings during the 1757 to 1857 period. It has an apathy and drift about the development of India. However, British Parliament has taken rights of ruling after 1857 independence war in India. The series of famine occurred during last quarter of 19th century raised the need of looking towards illness of agriculture in India. An establishment of famine commission (1880, 1898 and 1902) and irrigation commission (1903) had impact of intense discontent of Indians. These commissions had made constructive suggestions for improvement of agriculture in India. Therefore, some positive social and economic changes had happened during the British rule in India. The British rulers have made structural changes in land tenure system. The entire characteristics of land system had transformed by them through the introduction of British concept in India. The place of traditional share of government in farm produce has taken by the fixed revenue payment in cash. This system had not reference of good or bad harvest. The assessment of system and collection of revenue was individual basis or landlords appointed by the state in India, although it had varied according to circumstances of states and administrative convenience. Zamindari, Ryotwari and Mahalwari land tenure systems were existed in India. It was based on relationship between land holder and government. The nature of tenure in each case determines land revenue, gradation of interest, right of land, their recognition and interrelation and nature of assessment. An area under Ryotwari was 38 percent, 24 percent belongs to permanent zamindari settlement and 38 percent to temporary settlement at the time of independence in India. Major part of agriculture production was gone in the hands of zamindars and government in this land tenure system. It leads to tenants in vicious circle of poverty during British period. Initially, the western Yamuna and Eastern Yamuna canals were renovated and remodelled and reconstructed Kaveri Delta system. Similarly, Masonry weir across the Coleroon (1836), Upper Ganga canal (1842), Agra and Betwa Canals in Utter Pradesh, Sirhind Canal in Punjab, Mutha Canal in Bombay, Periyer canal in south India were the notable works consists to irrigation extension during British period. As well as, reforms made in banking was helpful for extension of banking sector during British period in India. The Co-operative Act (1904 and 1912), establishment of RBI (1935) and continuous assessment of banking sector had a positive impact on increasing quantity and quality of banks in India. Therefore, the average number of cooperative societies had increased up to 159180 and 107.32 lakh members during 1945-46 to 1949-50 in India. The British government had established research institutes and Agriculture Colleges for imparting agriculture education and research in India. Indian Council of Agricultural Research (1929), Indian Agricultural Research Institute, Delhi (1936), Indian Institute of Veterinary Research, Izatnagar, National Diary Research Institute, Karnat and many other research institutes were established for promoting research on various crops, fruits and animals during the British period. Almost, 95 agriculture colleges were established for providing theoretical and practical education of agriculture in India, which was connected with the Pusa Research Institute. It means the base of agriculture research and education had formulated by British in India. In short, the extension of irrigation, banking sector and agricultural research was the major positive works to develop agriculture sector. However, British had played role of administrator and collection of revenue than welfare of India. Hence, there was no motivation to increase crop yield. Consequently, agriculture sector ruined through exploitation and insecure land ownership in India.

Agricultural heritage and society

In India as in every country, agriculture was an integral part of popular background and gave rise to annual fairs, cattle melas, festivals and rituals, all of which were occasions for celebration. Almost every part of India had its own dates and customs for the purpose:

Akshaya Tritiya, for instance, a Hindu and Jain festival, is now often taken to be an auspicious day for buying gold, but it is also a harvest festival in parts of western and northern India. So are *Dol* or *Holi* (*Rang Utsav*) in the same region, Lohri in Punjab and neighbouring states, *Magh Bihu* in Assam, *Nabanna* in Bengal, *Onam* in Kerala or *Pongal* in Tamil Nadu, amongst others, most of them accompanied with rituals honouring cows and bullocks. Such festivals not only helped to bond local communities together, but have promoted national integration and continuation of agricultural heritage.

Many of you must have heard of *Kamadhenu,* the legendary cow of sage Vasishta which arose from the historic churning of the ocean which is described in the *Puranas.* The remarkable cow could mind-read the thoughts of the sage and instantly produce bountiful hospitality to any number of visitors to the ashram as continued symbol of agricultural heritage.

Perhaps the Aryan settlers had the legendary cow as a role model. They held the cow, bull and oxen in great reverence. "By means of cattle, soil is tilled and· production of food grains becomes possible. Oxen draw water, carry heavy burdens and draw carriages. Even if they are weary, they carry home the entire harvest.

The origin of Indian farmer's traditions and practices is traced to the scholars and sages who lived in the early Aryan settlements. These practices were needed once the pastoral groups settled down to regular cultivation of crops. These have been mentioned in the Rig Veda, the Atharva Veda, in the treatise by sage Kasyapa on 'agriculture, in the observations of scientists and physicians Parasara, Susruta, and Charaka, astronomers like Varahamihira and others. Scholars of different regions added to the store of knowledge from time to time. Many practices have been noted by Kautilya in the Arthashastra, written in about 250 A.D. Observations of foreign travellers in ancient and medieval India, have given detailed descriptions of agricultural practices and management.

Kaśyapiyakisisukti (800 CE)

This detailed dissertation gives advice on farm management not only to farmers but also to kings. The sages described the details about rice-growing practices are still widely followed in India today. 'Land is intended to receive excellence in every age.' 'A good quality land yields good results to everyone, confers good health on the entire family, and causes growth of money, cattle and grain.' 'To the west, north, east, or south of the villages and cities at the most convenient places, the king should prepare reservoirs of water according to the condition of the land.' 'The reservoir of water to be founded should be deep, equipped with barriers, splendid in the shape of a bow, long in some cases, round in others but essentially unfathomable.' 'They should also be equipped with inlets for water. Hence they should be founded near some hill or a high-level ground joined with a lake.' 'The king should plan its construction at such places as not to cause fear of danger from flooding. Such reservoirs should be regularly examined.' 'Large forests teeming with various trees, on the forest lands, or on the outskirts, or interiors of existing forests, or on mountain slopes should be propagated.'

About canals for irrigation, 'Even more than the ponds, lakes, wells, and protection of canals should be treated by them (farmers and the king) as their dharma, said the sages who know the truth.' 'That water therefore should be preserved by all sorts of efforts, as agriculture is said to depend on water. Hence, kings and other eminent persons should obtain water by exerting everywhere in the seasons and conserve it.' 'The second cultivation in a year is fruitful everywhere and is therefore recommended on various types of farmlands after rainy season rice for taking up this second operation, it is essential to raise the fertility of the soil, which can be achieved by using manure of goat-dung, cow dung, and vegetation (green manure).'

These are the messages of saints and sages to kings at every dynasty and king is being leader of region and of the community would come to follow.

Khana was a Bengali poet and legendary astrologer, who composed in the medieval Bengali proverbs between the 9th and 12th centuries AD. Her poetry, known as *Khanar Bachan* (or *vachan* i.e prediction) (meaning "khana's words"), amongst the earliest medieval compositions in Bengali literature, is known for its agricultural themes as folklore of forecast prophecy. '*Thakte balad kare naa chaass tar dukkho baro maas*' '*chase mulo tar ordhek tulo*'

'*Age dekho chashi bhai jeno hoy subhokhon, Shobhokhon dekhe soda koribe jatra, Pothe jeno na hoy oshubho barta*" Meaning: "The farmer should select an auspicious time before going to the filed with his plow. On the way he shouldn't listen to any bad news. This indicates the psychological condition of the farmer before going to work.

Another one: '*Mathe giye age koro dik nirupon, Purbo dik hote hal koroho chalon, Khana bole mor kotha shuno mohashoy, Phashal pholibe odhik nahi sonsoy*' and so on.

Daivajna Varāhamihir (505–587), also called Varaha or his son son Mihira, were Indian astronomer, mathematician, and astrologer born in Ujjain (or Bengal, according to some legends). The Indian Parliament building contains pictures of *Varahamihira and Aryabhata*, amongst other astronomers. Though little is known about his life, he supposedly hailed from South Bengal, where in the ruins of Chandraketugarh there is a mound called the mound of Khana and Mihir. Khana was the daughter-in-law of Varaha and wife of Mihir. Vararha was was astrologer King Vikramadiya and a famous astrologer herself. **Vikramaditya** was an emperor of ancient India according to oral traditions. Often characterized as an ideal king, he is known for his generosity, courage, and patronage of scholars. Vikramaditya is featured in hundreds of traditional Indian stories.

It is called that the predictions of Khana were so live and as they didn't dare to forecast those openly, that subsequently proved truth the rulers of said community gave her punishment to detach her tongue by knife to stop her forecast. These are also the continuation of agricultural heritage in the process of growth and development of agriculture and its advancement.

Conclusion

In India, agriculture started from old Babylonian days as 'creation of a Sumerian heritage'. A series of evidences explored elite antiquarianism in the Neo- and Late Babylonian periods available so far at the origins of a variety of disciplinary traditions in the ancient world, especially Greece and China as collected a series of essays on ancient canons and literary traditions. Present-day agriculture has changed all the ancient farming techniques replaced by modern mechanized farming around the world. But a growing sustainable agricultural movement, coupled with concerns about the impact of global warming, has led to a resurgence of interest in the processes and struggles of the original inventors and innovators of farming, these have got relevancy with agriculture developed since 10,000 to 12,000 years ago. The farmers cultivated crops and reared animals that grew and thrived in different diverse environments. In the process, agriculture developed adaptations to maintain soils, water and plant equilibrium keeping balance.

Some pertinent short questions

1. State few salient features of agriculture during British period:
2. What was Zamindari system in India?
3. Agriculture during vedic period;
4. Where the Pusa Research Institute was situated?
5. What is Cooperative act? When it was promulgated?

7

Heritage of Crop and Water Management in Ancient and Pre-historic and Medieval Period

Introduction

Apart from the general way of flood-flushing practiced at different sites of Indus Civilization, mention may be made of the following practices particularly undertaken in Sind. This is evident in the writings of inscriptions on seals from Mohenjo-Daro which state: 'Two canals were constructed in the month of *Kudam* (Aquarius) and *Mina* (pisces). Another seal records that it took 'thirteen months to construct a canal'. These canals were cut at low gradient and waters were allowed to merge ultimately into low ground in a direction oblique to the river so as to secure a great fall as possible. This fall varied from a foot to a few inches in a mile. They were up to hundred feet wide and about a dozen feet deep, not enough to draw off from the river except in flood. These were natural streams and probably seasonal. The perennial canals of the Indus system were cut to an extra depth and could receive water even when the main river was low, but their real purpose seems to increase the discharge during inundation.

Irrigation system

Irrigation system was consisting of lifting of water by water-wheel and by means of *Shaduf* from adjoining water sources. There was also provision of tank irrigation. Rice cultivation shows necessary moisture of soil probably retained by now using contour bunding for locking up monsoon rain-waters by tank-irrigation. Rainfall under vernal equinox or winter solstice has not been given much importance in the context of agriculture. Artificial rain making agents was the *twin stars, Asvinis* and *Dadhici,* described as horse headed sign. The exact time of their appearance cannot be ascertained definitely. Their association with cool season can be surmised from their name *Dasra* implying the same as referred to in the text of Panini. Moreover, their attributes as sender of *hima* (cold or snowfall) after the scorching heat of summer associates them as belonging to vernal equinox. That they appeared at the time of sun's movement to another house, is explicitly mentioned in the *Rigveda.* It is a fact which shows the rising of these stars at the junction of two solstices, summer and winter. The *Rigveda* in one place described them as associated with winter solstice. They were found to have been associated with three kinds of watery discharges, viz. rain *(vristi)*, dew *(madhu)* and snow *(hima).*

Drought and fog were considered the two enemies of crops. The drought was represented in the Rigvedaas the clouds (a) either not pouring down rain at all or (b) not pouring in time due to some astral connection or some other physical causes. The cloud comprised of: (i) the pre-rain drought-bearing agents, viz. *cumuri*and *dhuni;* and (ii) sterile dark (*Krishna*, *Strata cumulus);* (iii) cloud *(pipru)* causing much distress by oppressing heat, ultimately ended in rain (some form of *nimbus).* The impact of constellation upon drought was also mentioned in the *Rigveda*. The two constellations, *Vritra* (Hydra) and *Namuci* were associated with the non-falling of rain from cloud in summer solstice. Two types of drought were mentioned: (a) a severe and long continued form of drought ending in heavy shower; (b) a mild form of drought relieved by light showers. The term *kuyava* in the *Rigveda* signified fog appearing in the extremely cold winter night and continuing to the morning'. The term itself implies the phenomenon as harmful for barley crop.

The *Yajurvedic* period was marked by two factors: recognition of the constellation *krittika,* known in the Harappa period, as a multitude of seven stars which stand for seven rain-making phenomena in the firmament. Mention of twenty-seven constellations, some of which as their names suggested, threw light on different states of crops under the influence of these constellations. *Punarvasu* was related to sprouting up seeds of paddy or barley sown in the ground, Asresa (Hydrae) signified complete mature stage of crop. i.e., the grown-up plants of paddy and barley grow high enough to embrace each other, however, *Magha* (Regulus) indicated coming up of standing crop.

In the *Atharvavedic* period, the distinguishing factor in the idea of correlation between heavenly bodies and weather conditions laid in the recognition of healing virtues of the two asterisms, V*icrtau*in the constellation *Mula* and of the constellation *Rudra.* The former, not recognized in the *Rigveda,* were attributed with the power of curing diseases related to *ksetras*, implying, 'lingering diseases' and the 'diseases pertaining to field', i.e. soil-barrenness. The impact of the latter though recognized in the *Rigveda,* its special virtue in curing diseases of the cattle, was emphasized. The other astral influences on livestock and plant products comprised of: *Aryama* (the mid atmospheric condition between autumnal and vernal equinoxes) indicating increase of procreative power and growth of cattle; and *Asvinis:* Formation of sweet juice in plants and collection of the same by the bees. The importance of some astral influences during ploughing was found to have been established.

History of irrigation

Archaeological evidence of irrigation in Mesopotamia and Egyptian era as far back as the sixth millennium B.C.E., the cultivation of barley was mentioned under natural rainfall was insufficient to support such a crop. In the Zana Valley of the Andes Mountain in Peru, archaeologists found remains of three irrigation canals radiocarbon dated from the fourth millennium B.C.E., the third millennium B.C.E., and the ninth century C.E. These canals are the earliest record of irrigation in the New World. Traces of a canal possibly dating from the fifth millennium B.C.E. were found under the fourth millennium canal. Sophisticated irrigation and storage systems were developed by the Indus valley civilization in Pakistan and North India, including the reservoirs at Girnar in 3000 B.C.E. and an early canal irrigation system from circa 2600 B.C.E. Large scale agriculture was practiced and an intensive network of canals was used for the aim of irrigation. There's evidence of the traditional Egyptian king Amenemhet III within the twelfth dynasty about 1800 B.C.E., using the natural lake of the Faiyum Oasis as a reservoir to store surpluses of water to be used during the dry seasons, because the lake swelled annually as caused by the annual flooding of the Nile. The Qanats, developed in ancient Persia in about 800 B.C.E., are amongst the oldest known irrigation methods still in use today and hottest instrument is Persian wheel. They're now found in Asia, the center East and geographical region. The water distribution system emphasized including a network of vertical wells and gently sloping tunnels driven into the perimeters of cliffs and steep hills to tap groundwater. A water wheel with clay pots called noria round the rim powered by the flow of the stream was first use at about this point, by Romanians in geographic region. During 150 B.C.E., the pots were adjusted with valves to facilitate smoother filling as they were forced into the water.

The irrigation works of ancient Srilanka, the earliest dating from about 300 B.C.E., within the reign of King Pandukabhaya and under continuous development for the following thousand years, were one among the foremost complex irrigation systems of the traditional world. Additionally to underground canals, the Sinhalese were the primary to create completely artificial reservoirs to store water. The system was extensively restored and further extended during the reign of King ParakramaBahu (1153 – 1186 C.E.).

The oldest known hydraulic engineers of China were SunshuAo (sixth century B.C.E.) of the Spring & autumn period and XimenBao (fifth century B.C.E.) of the Warring States period, both of whom worked on large irrigation projects. within the Szechwan region belonging to the State of Qin of ancient China, the Dujiangyan Irrigation System was inbuilt 256 B.C.E. to irrigate an unlimited area of farmland that today still supplies water. By the primary century C.E., during the Han dynasty, the Chinese also used chain pumps that lifted water from lower elevation to higher elevation. These were powered by manual foot lever, hydraulic waterwheels, or rotating mechanical wheels pulled by oxen. The water was used for structure of providing water for urban residential quarters and palace gardens, but mostly for irrigation of farmland canals and channels within the fields.

In fifteenth-century Korea, the world's superiority gauge, woo ryanggyae (Korean), was discovered in 1441 C.E. The inventor was Jang Young Sil, a Korean engineer of the Choson Dynasty, under the active direction of the King, Se Jong. it absolutely was installed in irrigation tanks as a part of a nationwide system to live and collect rainfall for agricultural applications. With this instrument, planners and farmers could make better use of the data gathered within the survey. This study explores the conceptual and policy issues referring to the impact that irrigation has on crop production, farm income, inequities in income distribution and poverty alleviation. It also focuses, specifically, on poverty issues related to head-tail water distribution inequity in an irrigation system. Improved access to irrigation infrastructure will increase crop yield, agricultural production and farm income within a section. However, the character and scale of feedback effects related to irrigation access and their impacts on farm income and poverty reduction process aren't yet clearly understood or reported in irrigation literature. The literature has focused more on the direct impacts of irrigation, but not on indirect impacts like rural employment and economic multipliers related to the availability of irrigation during a region. Likewise, a touch explored topic is how exactly irrigation infrastructure affects inequalities in income distribution and also the poverty status of a locality. Equally, the feedback effects and therefore the nature of secondary benefits related to this process also are unexplored. Within the context of decreasing planet market food prices over the last 20 years, the secondary benefits of irrigation provision are increasingly receiving central focus in irrigation investment decision making. In fact, these secondary benefits are now considered more important than the direct benefits of an irrigation project. From the synthesis of results presented in contemporary irrigation literature and also the evidence provided during this paper, we conclude that improved irrigation access could be a powerful instrument for reducing rural poverty during a given region. This is often not such a lot through the direct impact of increased yield and farm returns intrinsically, but more through indirect impacts like increased rural employment and therefore the feedback and multiplier effects related to the supply of irrigation infrastructures. the dimensions of multiplier effects generated in an irrigated agricultural society depends upon the amount of backward and forward linkages operating within the particular regional economy, i.e., the character and scale of inter linkage within and among rural enterprises and also the inter linkage with market infrastructure. In other words, increased crop yield could be a necessary condition but not the sufficient condition for poverty alleviation in an exceedingly region. The impact of irrigation on poverty reduction depends upon the structure of a rural economy and on how the extra farm income generated by improved access to irrigation is really spent within a rural economy, and its feedback impacts on rural employment and rural wage structures. Therefore, the extent of economic multipliers operating in an exceedingly regional economy is crucial in determining the impact of irrigation on the poverty status and therefore the inequalities in income distribution within that individual economy. The main focus of policy-makers within the irrigation sector has now shifted to issues like irrigation water management, participatory decision-making and institutional reform within the irrigation sector, environment management for system sustainability and more equitable distribution of advantages across irrigation systems and across agro-environments. Of these changes are visible in efforts at reducing the extent of governmental failures and market failures related to managing irrigation commands. The underperformance of canal systems has further aggravated the income gap and also the relative personal income within irrigation systems, resulting in an unequal distribution of irrigation benefits across sub-systems. In reality, the wealth creation and trickledown effect aimed toward alleviating poverty in irrigated areas isn't happening within the originally envisaged manner. Therefore, additional direct public policy interventions and more pro-poor institutional and policy reforms v are required. This can help minimize the differential distribution of advantages across sub-systems and farmers, and increase the social benefits and well-being provided by the supply of irrigation infrastructure. Ultimately, irrigation could be a typical public good, either directly provided or largely subsidized by governments for overall social well-being. The restructuring of irrigation commands can be achieved through reforming of institutional, technical, managerial and operational factors. A number of the institutional reforms are: improved stakeholder participation in resource use decision-making, participatory irrigation management (PIM), and irrigation management transfer (IMT), defining clear water rights and water entitlements and self-enforcement of efficient charge collection mechanisms. A number of the technical factors for improving water allocation across sub-systems are better water control structures, laser leveling, lining of canals, improved water storage systems, conjunctive use of rain, canal and groundwater, etc. Likewise, a number of the managerial and operational factors involved here are: better enforcement of existing rules and regulations to attenuate the lawlessness seen in irrigation commands, improved operation of systems, tailoring the irrigation operation and maintenance costs supported incremental benefits generated and level of water uses within the system, targeted additional financial and credit interventions within the system considering the necessity of tail-end farmers for extra irrigation equipment, and improving field structures and water storage.

Hence, Irrigation is the application of controlled amounts of water to plants at needed intervals. Irrigation helps to grow agricultural crops, maintain landscapes, and revegetate disturbed soils in dry areas and during periods of less than average rainfall. Irrigation also has other uses in crop production, including frost protection, suppressing weed growth in grain fields and preventing soil consolidation. In contrast, agriculture that relies only on direct rainfall is referred to as rain-fed or dry land farming.Irrigation systems are also used for cooling livestock, dust suppression, disposal of sewage, and in mining. Irrigation is often studied together with drainage, which is the removal of surface and sub-surface water from a given area.

Use of irrigation water

Irrigation was used as a means of manipulation of water in the alluvial plains of the Indus valley, the application of it is estimated to have begun around 4500 BC and drastically increased the size and prosperity of their agricultural settlements. Sophisticated irrigation and water storage systems were developed by the Indus Valley Civilization, including artificial water storage at Girnar dated to 3000 BCE and an early canal irrigation system from 2600 BCE. Large scale agriculture was practiced and an extensive network of canals was used for the purpose of irrigation. Perennial irrigation was practiced in the Mesopotamian plain whereby crops were regularly watered throughout the growing season by coaxing water through a matrix of small channels formed in the field. Ancient Egyptian practiced basin irrigation using the flood by inundated of Nile River which had been surrounded by dykes. The flood water was held until the fertile sediment had settled before the surplus was returned to the watercourse. There is evidence of the ancient Egyptian king Amenemhet III in the 12th dynasty in 1800 BCE using the natural lake of the Faiyum Oasis as a reservoir to store surpluses of water for use during the dry seasons. The lake swelled annually from flooding of the Nile.

Since the time of old Mughal period water resource development and its management was in remarkable manner but in later era of Mughal Emperor Bahadur Shah II it was redefined to store water and development. The *sakia* is the water distribution wheel used by Ancient Nubians for agricultural uses. Agricultural irrigation system was developed by Nubian in transition period of 2nd and 3rd millennium BCE. The water resource is depend upon the River of

Nile and others river of Sudan. The sub-Saharan Africa irrigation met up by the River Niger and culture and civilization was developed during 1st millennium BCE.

Terrace irrigation is evidenced in pre-Columbian America, early Syria, India, and China. In the Zana Valley of the Andes Mountain in Peru, South America, and archaeologists found remains of three irrigation canals during 4th millennium BCE. These canals are the earliest record of irrigation in the New World with new routine direction. Traces of a canal possibly dating from the fifth millennium BCE also canal irrigation was found under the 4th millennium. Iranian irrigation was remarkable during 6th millennium BCE where barley crop was major priority and areas where natural rainfall was insufficient. The Qanats was developed in ancient Iran in about 800 BCE, are among the oldest known irrigation methods still in use today. They are now found in Asia, the Middle East and North Africa. The system comprises a network of vertical wells and gently sloping tunnels driven into the sides of cliffs and steep hills to tap groundwater. The popularly known noria, a water wheel with clay pots around the rim powered by the flow of the stream, was first brought into use at about this time by Romanian residents in North Africa. By 150 BCE the pots were fitted with valves to allow smoother filling as they were forced into the water.

The irrigation works of ancient Sri Lanka, the earliest dating from about 300 BCE, in the reign of King Padukabhaya where it was under continuous development for the next thousand years, were one of the most complex irrigation systems of the ancient world. In addition to underground canals, the Sinhalese were the first to build completely artificial reservoirs to store water. These reservoirs and canal systems were used primarily to irrigate rice fields which require a lot of water to cultivate. Most of these irrigation systems still exist undamaged up to now, in Anuradhapuram and Polunnaruwa, because of the advanced and precise engineering. The system was extensively restored and further extended during the reign of Emperor Parakram Bahu during 1153 to 1186 CE.

Irrigation has been practiced since time immemorial, and it was started but evidences say that it is the foundation for all civilization since great civilization were started in the river basins of Sind and Nile. This civilization was developed with when need to irrigate the crop and system failed to maintain crop production. There are some evidences that during the Vedic period (400 B.C) people used to irrigate their crops with dug well water. Irrigation was gradually developed and extended during the Hindus, Muslims and British periods in Indian continents. The Grand Anaicat constructed across the river Cauvery is an outstanding example for the irrigation work by a Chola king the great Karikala Cholan during second century was remarkable irrigation era in agriculture sector. The Veeranarayanan Tank and Gangai Konda Cholapuram tank was constructed during 10th century in Tamil Nadu and Anantaraja Sagar in Andhra Pradesh was constructed during 13th century.

The remarkable construction of wells and tanks was observed by Mauryan king Samudragupta and Ashoka. The human being needs their food in every time and it is interrelated with water, without water crop production not possible as because later Moghul kings or North India and Hindu kings of South India focused their attention, in the establishment of canals, dams, tanks etc. Gradually, it was also seen by British Government, initiated their work during 19th century in remodeling and renovation of the existing irrigation system. The Upper Ganga canal, Krishna and Godaveri delta system, Mettur and Periyar dams are the great irrigation structures built by the British rulers. After independence, Irrigation activities have been accelerated and number of multipurpose river valley projects like Bhakra-nangal in Punjab, Tungabhadra in Andraprdesh, and Damodar Valley in Madya Pradesh was established time to time as per need of irrigation to the crop.

Objectives of irrigation application

Irrigation is practiced to maintain the different developmental parameters. Those are:

a) To make up for the soil moisture deficit;

b) To ensure a proper & sustained growth of crops; to make harvest safe;

c) To colonize the cultivable wasteland for horizontal expansion of cultivation;

d) To shift from seasonal cultivation;

e) To promote more intensive cultivation by multiple cropping;

f) To improve the level of agricultural productivity by acting as an agent for adoption of modern technology;

g) To lessen the regional & size-class inequalities in agricultural productivity that will reduce in turn socio-economic imbalances.

The effect of supplemental water application as irrigation for crop production at its needs during rainless period is well established throughout the world. The plant has life, so it requires water for growth and development to give us proper quality and good yield. Water is required for metabolic activities including solubilization, absorption and translocation on the way of biochemical processes in photosynthesis, transpiration and respiration. The objective of irrigation management is different where availability of water is abundant that of limited or restricted water supply. The former is associated with yield increase per unit of land whereas later is to increase crop yield per unit of water use. Water requirement of crops response to water use vary with the different growth stages of the crop. The certain physiological growth stages are more sensitive to water shortage. So particularly where the water supply is limited, it is necessary to take into account the physiological critical growth stages with respect to moisture requirement to get more water use efficiency. If the crop suffers from water shortage to such growth stages, either quality or yield of the crop reduces drastically. Such as in case of maize: vegetative, tasselling and silking at about 20-40 and 50-60 days, respectively, after sowing have been observed as critical in the demand for water. In case of pulses including beans profit comes from irrigation during flowering and podding; as in case of oilseeds, the flowering and fruiting are considered as critical growth stages. If wheat crop suffers from water shortage at crown root initiation stage (21 days after sowing), it will hamper the tillers production, which would give poor quality and yield of crop. So the optimum moisture conditions in the soil profiles are very essential to maintain qualitative and quantitative standard of the crop.

Rivers and water resources management

Water management measures comprised diverse ways of dispersal of water to the fields; conservation of water and protection of soil generally developed in the regions of scanty water supply or where water-resources were subject to depend on flood-water or rain water in matters of increase and decrease of its volume. *Saraswati* the non-perennial rain-fed river having risen in dunes or valleys of *Siwalik* range known as lower Himalayas during Holocene period underwent several changes in its entire course from emergence to fall in Arabian Sea. The river had been identified with modern *Sarsuti* which took its rise in the *Sirmur* hills of the Himalayan range, flows past Ambala, Pipli, Kuruksetra, Pehowa, Sirsa and ultimately lost in the desert of Bhatner. The river was different names in its different courses, viz. *Ghaggar* from its union with the same while flowing in Rajasthan, *Hakra,* in its flow in the east of Indus and *Nara* in its converge with Indus drainage. The appellation *Saraswati* was attributed by the Vedic sages to distinguish it as a 'river full of lakes'. The uppermost form of the river is still known by this name. The *Mahabharata,* the *Puranic* epic also stated a number of names of Saraswati in its flows in different directions. Saraswati as described in ancient texts particularly in the Rigveda and *Mahabharata* was rich in effluents and distinguished by its flow in different forms and in different courses even sometimes in sub-terranean regions. The Rigved are ferred to 14th main effluents and tributaries of Saraswati. Seven amongst them were associated streams, of which three were original, formed higher up with the original stream. The other four were the branches of westernmost of them. These three original forms were known as eastern Saraswati, middle Saraswati and western Saraswati. The *Atharvaveda* named these three Saraswatis and later Panini described their courses. The other seven were sister rivers falling in Saraswati basin. These were Drisadvati, Apaya, Hariyupiya, Yavayavati, Sarayu, Kulsi and Virapatni. In addition to these, mention was made of numerous lakes in Vedic literature where Saraswati was dispersed naturally at the time of flood. The *Mahabharata* referred to seven Saraswatis, indicating seven branches of the river. The valley below Pehowa was known as *Sapta Saraswat,* i.e. the place where the river divided itself into seven streams. These seven seem to be canal streams.

Saraswati as the later Vedic texts and the *Mahabharata* stated, disappeared in the desert at Vinasana before its meeting with Indus drainage. Its reappearance took place at Camasodbheda. The final union of Saraswati with sea was specifically mentioned in the Rigveda and in the *Mahabharata.* Two sites of this union were recognized in the *Mahabharata* one is Otu in Hisar District on the border of the desert; the other at Prabhasa. The entire course of Saraswati thus appeared to have covered Punjab, northern and western Rajasthan, Sind and Surashtra.

The several sources of water for irrigation were rain water and draining of water from the rivers. Saraswati River by its rich water resources is found to have sustained to great extent irrigation in a wide area of north-western India from Punjab to Sind including Rajasthan and Surashtra while traversing the regions in course of its different flows. Its contribution to irrigation particularly in Rajasthan desert area during 2nd and 3rd millennium B.C. was worthy to note. The Rigved are referred to Saraswati as the conjunction of five tribes and described its water possessing high fertility capacity. There were two fold contributions of river Saraswati to irrigation by surface flow, and sub soil flow of water.

Surface flow resulted in formation of lakes and pools due to high spate in river after torrential rain. This helped draining the land of a larger area and ensuring cultivation. Surface flow also ensured tapping of water through canal. However, the sub-soil flow was a perennial source of water to dug-well. In all probability well-irrigation was widely practiced in Rajasthan due to availability of pockets of such water under the sandy ground. The extinct of the river from several regions appeared to have been caused from the following factors such as stoppage of supply water from other perennial flows and clearance of forest and excavation of canals for different purposes.

River water was tapped for irrigation purpose where there were seven principal rivers by which India was known as the "land of seven rivers" enumerated differently by different authorities. These are Ganga, Yamuna, Godavari, Saraswati, Narmada, Sindhu and Kaveri a group representing rivers of the entire Indian sub-continent. Further, Panini referred to tapping of water from rivers like, Sindhu, Suvastu, Varnu, Sarayu, Vipasah, Chandrabhaga and Devika. Two ways of direct supply of river water were flood-flushing and over-flow irrigation system. The latter was particular only to Bengal and later in South India.

River associated streamlets serving the purpose of irrigation

(a) Natural water course

Inundation Lake such as Mancher associated with Indus. The water is believed to have been passed to the lake through Nara.

(b) Artificial water courses

The water courses which is associated with two rivers of western India, canal, nullah, etc

The associated factors promoted irrigation, is found to consist of re-opening of rivers at their heads closed from natural causes, the clearance and excavation of dead water courses and minor channels, flood-flushing under control through the construction of dam, canals and sluices in embankments.

The rigveda were two types of irrigation systems were practiced during the 1500 BCE (i) Natural irrigation by river water distributed over arable field by means of channels. *(ii)* Artificial irrigation by wells when the water was lifted by means of stone-wheel, *ghati-cakra*or *ghati-yantra.* The later records described, these tools consisted of a row of earthen pot tied to the rim of a drum shaped wheel turning into a vertical plane over water. The wheel, having spokes, was worked like a capstan. It was moved by gearing a horizontal wheel turned by man or beast to the vertical water hoist. *Yajurveda* (1200 B.C.-1000 B.C.) referred irrigation by dam or reservoir and by canal. *Atharvaveda* (c. 1000 B.C.) narrated about canal irrigation. The other texts of the Vedic literature referred to large-scale use of irrigation by well and reservoir dam.

The productivity of a piece of land, as conceived by the ancient Indian economist, depends on four factors, viz. extension, size, situation and accessibility as addition to soil, irrigation facility and productivity of land. As per Vedic texts, 'extension' means 'utilization of different sites for cultivation of crops other than the grain crops'; 'size' indicates 'large and small holdings'; 'situation' stands for 'particular region like, *jangala. anupa*and *sadharana*, where arable plot is situated; and 'accessibility' suggests proximity to market. Ideas on different aspects of farm land as deduced from relevant sources.

Existence of different types of agricultural holdings was assessed in an agricultural site at Inamgaon in Western India. The smallest one holding was of hundreds square mile and the biggest one was of several hectares. During the later stage, the appearance of *grama*in the sense of estates came into existence in the *Atharvavedic* period. In the later period revenue was imposed on the value of the estates determined from its resources. From Sutra period onwards there appeared land measures, viz. *nivartana*, which, according to later interpretation indicates a piece of land sufficient for the subsistence of one person by the agricultural outturn from the field.

Irrigation projects

There was a time when the multipurpose river valley projects were hailed as the harbingers of prosperity. These "temples of Modern India" as Jawaharlal Nehru termed the dams, have no doubt proved an important factor in achieving a successful breakthrough in agriculture and a measure of self-sufficiency in food production.

However, their success has been overrated. Serious problems have arisen. Heavy siltation of the reservoirs of major dams reduces storage capacity, and they become incapable of absorbing heavy floods. Disastrous fallout is the degradation of the soil in the command areas of irrigation projects due to water logging and soil salinity.

Canal irrigation, again, has contributed to land degradation. Groundwater table rises due to accession of applied water to groundwater through deep percolation form unlined field channels, distributaries and main canals, and restrictive drainage due to construction activities.

When the groundwater table rises to within 1 to 2 meters of the ground surface, it contributes significantly to evaporation form the soil surface, leaving the dissolved salts in the root zone of soils. The salt content of most groundwater in the arid and semi-arid regions is high. This further accentuates the processes leading to the formation of salt-affected soils. There are numerous examples of land degradation following the introduction of canal irrigation.

The water from the Bhakra canal system was made available in Hissar district in Haryana in 1963. Large areas irrigated by the carnal are already salinised. Similar experiences are available in the area of the Tungabhadra irrigation project in Kamataka and in the NagaijunaSagar command area. The problems in the Rajasthan Canal command area have acquired serious proportions.

Efficient water use

The emphasis on minor irrigation works in recent years is worth noting. Groundwater development is part of this. Minor irrigation schemes, besides being an instant and reliable source of water, also help to control water logging and salinisation in canal command. States have taken up 'on-farm development' works in all the command areas and are introducing warabandhi water delivery methods to reduce water application even while increasing farm production.

Sprinkler irrigation is water efficient, and has been introduced in canal irrigated areas of Haryana, Rajasthan and Madhya Pradesh. Drip irrigation is being encouraged in Maharashtra, Karnataka, Andhra Pradesh, Orissa, and Tamil Nadu for fruit crops. Fertigation is an application of fertilizer through drip irrigation to maximize farm productivity with available water.

As for surface irrigation, scientists have established that paddy fields do well with a much lower depth of standing water than commonly in practice. Methods to conserve water include intermittent water delivery, furrow method, micro-irrigation in well- irrigated areas, and delivery of water in pipelines to avoid loss on way. The green-house method is another scientific way of growing special crops.

Irrigated agriculture

The effect of supplemental water application as irrigation for crop production as it needs during rainless period is well established throughout the World. The plant has life, so it requires water for growth and development to give us proper quality and good yield. Water is required for metabolic activities including solublization, absorption and translocation on the way of bio-chemical processes in photosynthesis, respiration and transpiration. The objective of irrigation management is different where availability of water is abundant that of limited or restricted water supply. The former is associated with yield increase per unit of land, irrespective of amount of water used for the purpose, where, later is to increase crop yield per unit of water use. Water requirement of crops and crop response to water use vary with different physiological growth stages. The certain growth stages are more sensitive to water stress situation. So particularly, where water supply is scarce or limited, it is necessary to take account of application of water to the critical growth stages with respect to moisture requirement to get higher water use efficiency. If the crop suffers from water shortage during this period, either quality or yield of the crop reduces drastically. The optimum moisture conditions in the soil profiles are very much essential to maintain for getting the qualitative and quantitative standard of the crop output.

Efficient water management technologies

All Indian Coordinated Research Projects (AICRPs) have developed technologies focusing on enhancing water use efficiency at all levels. They are briefly described below.

a) Land leveling

Proper land leveling is one of the management options which is generally ignored. It increases the water application efficiency which leads to higher yields as well as increase in water use efficiency. It also has a direct impact on the nutrient use efficiency.

b) Methods of irrigation

Check basin and border strip irrigation: Over the years many surface methods of irrigation have been developed. Among them, the check basin method of irrigation is the most popular. It is very important to employ the correct method of water application to minimize the adverse effects of irrigation. The selection of the right method of irrigation is influenced by the soil type, land topography, crops to be grown, quality and quantity of water available for irrigation and other site-specific variations. It is essential to follow the prescribed guidelines to improve productivity per unit quantity of water applied. Considerable amount of work has been done on these aspects but the guidelines for selecting the appropriate method of irrigation water application are largely ignored. Farmers go in for surface flooding which is an inefficient manner of using this precious natural resource. 10 to 15% efficiency in water use can easily be achieved by adopting the appropriate method.

c) Furrow irrigation

The furrow method of irrigation is generally used to irrigate row crops and vegetables, and is suited to soils in which the infiltration rates are between 0.5 and 2.5 cm hr-1. It is ideal for slopes varying from 0.2 to 0.5% and a stream size of 1-2. Many of the field crops in which water is applied through flooding, check basin or border strip methods, can easily be adapted for furrow irrigation or its modified version i.e. raised bed system. Table 5 shows how savings in irrigation water can be achieved by switching over to raised bed furrow irrigation systems. In areas where water for irrigation purposes is scarce, the practice of alternate or skip furrow irrigation can save considerable quantity of water without significantly affecting the yields. In areas requiring surface drainage or where the cultivated crops are sensitive to water logging, the furrows are effective in removal of excess water.

d) Surge flow irrigation

Excessive water intake and deep percolation losses are major limitations for irrigation through furrows and border strips. Surge flow irrigation, the intermittent application of water in a series of on and off modes of constant or variable time spans has the potential of reducing intake and percolation losses, increasing the irrigation efficiencies and conserving irrigation water.

Experiments conducted at several locations have corroborated that surge flow furrow irrigation results in faster water front advance and reduces the volume of water required to complete the advance phase. The studies also confirmed that it increases water distribution efficiency, reduces infiltration rate and the cumulative infiltration depth, and consequently reduces the percolation losses.

c) Pressurized irrigation systems

To encourage adoption of water efficient technologies, the Government has provided subsidy to popularize pressurized system of irrigation like sprinklers and drip or Trickle irrigation which both together now occupy more than 6 million hectare. The pressurized method of irrigation gives many advantages over the gravity surface irrigation methods in terms of water savings and yields. The quality of produce through drip irrigation is generally superior to conventionally irrigated crops. These pressurized irrigation systems are very useful particularly in undulating and uneven lands or very coarse textured soils. The use of these techniques can be promoted for use in irrigation commands for specific conditions. Pressurized system of irrigation popularly called as fertigation.

Use of poor and marginal quality water for irrigation

Agriculture will also have to depend more and more on the use of marginal/ poor quality of water as it is projected due to intense competition with domestic, industrial and power sectors that the share of water allocated to agriculture is likely to decrease by 10-15% in the next 20-30 years. Sprinkler and drip irrigation methods of water application are techniques in which such water can be used effectively. However, the long-term impact of prolonged use of such waters has to be always kept in mind while recommending such In regions where good quality water is available even in limited quantities, conjunctive use offers a safe way of utilizing a combination of poor and good quality either through mixing. In a detailed study aimed at evaluating the use of paper mill effluent for irrigation in the farmers' fields, it was observed that in all the crops namely, maize, wheat, rice and mustard, highest grain yields were obtained with mixed irrigation.

The water savings technologies developed

a) Proper scheduling of canals (matching supply with demand): 40-60%

b) Precision leveling through laser levelers: 15-20%

c) Scientifically designed check basins/border strips: 10-30%

d) Zero tillage: 20-30%

e) Adoption of pressurized irrigation systems: 40-70%

f) Land configuration changes-ridge/furrow or raised/sunken beds: 20-25%

g) Use of tissue cultured eucalyptus, banana, sugarcane, papaya etc. 30%

Water management

There is an apparent ample calibration irrigational practices, some of which appeared to accept been in accepted conveyance and some accessories were regional. These comprised of the following:

Flood water into basin concept to authority of water for the purpose of irrigation comes into force. The basin Sudarsana was discovered by Asoka to facilitate irrigation. It was affiliated with two rivers of western India. In case of aqueduct irrigation in which the basic aqueduct is declared to accept been provided with two sub-canals, one confined as agriculturalist and the added as supplier. Ascendancy of water by architecture of sluices threw ablaze on management of aqueduct water for irrigation.

Well-irrigation was empiric during indigenous and ancient age. This consisted of ample calibration well-irrigation with the actualization of two types of well, namely, ring able-bodied and absorb able-bodied appear from altered sites of excavation. Panini as well referred to two types of well, Karkandhu and Sakandhu. The closing was declared to be acclimated by the Sakas and it denoted an anatomy of water wheel.

Canal irrigation was as well acclaimed by historians. The canals were kept abounding either by hand with the aid of water skins or antithesis pole, or by water transported, on the backs of animals, or by application a bucket chain. An able adjustment was begin to accept been formed out by oxen aggressive up a acclaim angled bogus backbone and bottomward it afresh time afterwards time, in accomplishing so carriage up from a able-bodied on anniversary break a leather-bucket abounding with water emptied afresh into a accumulation canal. Water-lifting accessories were comprised of automated accessories by application lever, caster and axle, and application of power. Analytical adjustment of channels for arrival and address of water from the acreage was fabricated to adulterate the crop. Attention of rainwater in channels and its management in paddy fields were as well done.

Conclusion

Now-a-day, a number of water management technologies for field crops have been developed over years to enhance agricultural production. The area of crop(s) under irrigation was also increased substantially. The irrigation potential was created at the cost of huge amount of national exchequer. But there is still a large gap between the irrigation potential created and utilized.

The water management technologies developed so far was not also being practiced at the actual users' end. There are gaps in the existing system, which acts as constraints on adoption of water management technologies.

To address the problems and to bridge such gaps, the issues and priorities of water management should have major influence on the end users levels on adoption of water-efficient management practices for growing field crop(s).

Some pertinent short questions

1. What is the relation between irrigation and poverty alleviation?
2. What is relation between irrigation and civilization?
3. What do you mean by poor and marginal quality of water?
4. What is efficient water management technology?
5. What do you mean by canal irrigation?

8

Journey of Indian Agriculture and Its Development from Past to Modern Era

Introduction

Indian agriculture began by 9000 BCE as an after effect of earlier agriculture of plants, and conduct of crops and animals. Settled activity once again followed with accessories and techniques accepting developed for agricultural heritage system. Previously, double monsoons led to two harvests accepting reaped in one year alike various countries. Indian online inscription once again able the prime via complete trading networks and adopted crops were at variance in India. Plants and animals are advised basal to their adjustment by the Indian that came to be admired and venerated. The boilerplate ages are of saying irrigation channels adeptness a new affiliated of accord in India and Indian crops affecting the economies of added regions of the world. Land and water management systems were developed with an aim of accouterment accordant growth and crop productivity. Despite some stagnation during the afterwards strike era the Republic of India was able to beforehand a complete agricultural programme. In India, there are about 110 crore of farmers engage i.e. two-third of complete workforce, contributes to 26 per cent of the GDP and about one-sixth of the complete address earnings. India associated a repulsive agricultural system at the time of expertise growing at a ritual quantity in advance of 0.3 per cent in the earlier bisected of the stand for the century. Perplexing building up of lateral population pressure provided by agricultural scientists and harder plan put in by the farmers due to the planned Government agricultural policies. Indian agriculture has able to provide a service in advance an amount of 2.7 per cent in the procession expertise in the following era. The beforehand in food particle accession was a lot of arresting at about 3.7 per cent compared to population before of 2.2 per cent during the decade of 1980's. Unfortunately, it slowed down in 1990s to 1.8 per cent, lower than the population earlier by 2 per cent. Further, rural agriculture across too, the development has been admit in accession and prosperity of crops in several regions. All said and done, over-riding the three-fold in advance at population by a four time admission in food stuff succession from the earlier 5 year plan to the change about of the period by all agencies an admirable achievement. The key macro indicator of borough expertise in agriculture comprised in advance jump in food succession from budgetary provision of rupees 50.8 crore in 1950-51 to 212 crore in 2000-2001. It is annual budget up to date that 92 per cent of the entrance in achievement has been technology-led. Thanks to the agricultural scientists for developing location based technologies and animate off to the Indian farmers for their activity to ideas, activity to acclimatize and as able to accepting of new and bigger authentic knowledge.

Planning Commission's incorporation under unrest situation of the country that the embracing beforehand volume of crop adoption below from 3.72 per cent per annum in 1980s to 2.29 per cent in 1990s. Similarly, prosperity fell down from 2.99 per cent per annum to 1.2 per cent. Analyzing discarded food and non-food grain crops, it is credible that the deceleration in the beforehand an amount of food grain was abrupt and it fell from 4.02 per cent to 2.83 per cent during that period. This apart, the prosperity of crop fastens abides to comfortable lower compared to put in crop yield. This accepts to draw moderate incorporation now. It may be annual direction that during the stand for three decades, the attainment entrance was ushered by use of top acknowledging quality varieties of seed, assured accession of water and accelerated use of fertilizers and pesticides. These acquire their limitations and problems. For example, calculations actualization if the acclimatized in advance beforehand volume in agriculture of about 4 per cent is to be able to increase through the currently acclimatized technology mixed with the per hectare fertilizer use that acquires to admission to 186 kg/ha in 2011-12 from the used to amount average of 76 kg/ha earlier. Such an allied of fertilizer activity may be repulsive for both atmosphere and prosperity reasons. Biotechnology or supplementary technologies could be an acceptable at very small allocation of the reception to the problems. These are to be addressed adequately. The millions acquiescent for low prosperity could in large quantities be, appraisal in advance belted few crops growing in several agro-climatic zones including rainfed crops, low fertilizer use and as low fertilizer use efficiency, short form of food giving out and volume succession to agricultural and vegetable crops, top post-harvest losses strangely in agricultural after-effects (20-40 per cent), complete low agricultural consequences; poor exchange infrastructure, flourishing terrible proficiency in agriculture across unusually as entrepreneurs and skill workers. Special efforts allegation to be false to succession the prosperity and attainment of crops, livestock, fisheries, and agricultural crops to ensure food and comestible advocacy for the people and descent surplus for address befitting in actualization the accomplishment that India has an affluently advancing advantage in agricultural export. Systematic and accurate investigations in prime crops such as rice, wheat, maize and others accept arise that plentiful advice in this attention wherein it can be deduced from archaeobotanical, archeological and mysterious annual sources.

Indices to analyze age-old agriculture

The harder abstracts from the subsequently available sources accept been acclimated to draw inferences apropos introductions of crop breed during age-old times;

- Archeological remains: Sculptural and clear delineation of plants or a rhizome share in architectural monuments, such as caves, temples, etc. anachronous to age-old times;
- Archaeobotanical remains: The charcoal of corm actual acquired from various archeological blasting sites with carbon dating for age-old period;
- Literary sources: Advertence of tuber breeds in writings/literature of oriental language, completely the Sanskrit, the Vedic conceptual with colloquial name; and
- Ancient dating: Based on the evidence, advertisement of those plants falling into the period above-mentioned to 8th medieval period. Archeological perfect confirmation remained arch of crop corm agriculture those are food in attributes in age-old times, absolutely from Neolithic-Chalcolithic period in Harappa, Gangetic Plains, and other additional locations of the country including South India. They prove agnosticism the presence/cultivation, thereby age-old accession of crops; complete the controversies apropos crops such as maize, sunflower, cashew nut, tamarind, and others.

However, in afterward with above criteria, the difficulties listed below are acceptable to be encountered.

- Limited archeological affirmation in crops broadcast by abundant bureau (leafy vegetables, basis and bake-apple crops);
- Lack of expertise of oriental languages, such as Sanskrit, Prakrits, Pali, Dravidian, etc.;
- Difficulties in identification and appraisal of Latin synonyms of rhizome species;
- Poor documentation/inadequate advice apropos accessory crops of neighboring significance;
- Difficulties in appropriate the age-old introductions from those from medieval period because of synopsis of earlier advice and pro-west

attitude of historians. In crops of African origin, admission to the Indian Subcontinent, both through land and sea-route, is possible;

- It is difficult to analyze in the middle of the two routes and in some crops, it ability accept occurred through both. Phylogenetic studies including use of granitic markers may be bare to judge on the adjoining dates and relationships with complete biogenetic diversity and

Linguistic abstracts accept hardly been used; however, in the present commodity advertence to Sanskrit writings and Sanskrit names and their affinity with the names in their country of agent accept as well been considered, absolutely in case of American and African crops. In this regard, Indian description *Bhava Prakash Nighantu* (Indian Materia Medica) with footnote in Hindi by Bhava Mishra (c. 1600 AD) has been relied upon. The age-old evidences arresting accession and conduct of altered crops are supplementary corners of the world.

Chronological agricultural technology development in India

Agriculture in India is broadly classified into five different periods before India's independence.

1. Early history (Before 15000 BCE)
2. Vedic period – Post Maha Janapadas period (1500 BCE – 200 CE)
3. Early Common Era – High Middle Ages (200–1200 CE)
4. Late Middle Ages – Early Modern Era (1200–1757 CE)
5. Colonial British Era (1757–1947 CE)
6. Republic of India (1947 CE onwards)

A. **Early history (Before 1500 BCE)**

12000 to 9500 years ago

i. Hunters and food-gathers stage existed.
ii. Stone implements were seen throughout the Indian subcontinent.
iii. Domestication of dog occurred in Iraq.
iv. Earliest agriculture was by vegetative propagation

9500 to 7500 years ago

i Wild ancestors of wheat and barley, goat, sheep, pig, and cattle were found.

ii. 9000 BCE: Wheat and barley were domesticated in the Indian sub-continent. This period also saw the first domestication of the elephant in addition of domestication of horse, sheep and goat.

iii. 8000-6000 BCE: Wheat and Barley cultivation, domestication of cattle, primarily sheep and goat was visible in Mehrgarh (Balochistan, now in Pakistan). In India, agricultural activities included threshing, row planting - either of two or of six, and storing grain in granaries.

7500 to 5000years ago

i. Significant features were invention of plough, irrigated farming, and use of wheel, and metallurgy and in Egypt, seed dibbling.

ii. 5440 BCE: Wild Oryza rice appeared in the Belan and Ganges valley regions of northern India. Rice cultivation was seen in the Indus Valley civilization.

iii. 5000 BCE: Agricultural communities became widespread in Kashmir.

5000to 4000 years ago

i. Harappan culture is characterized by cultivation of wheat, barley and cotton; plough agriculture and bullocks for drought.

ii. Wheeled carts were commonly used in the Indus valley.

iii. Harappans not only grew cotton but also devised methods for ginning / spinning / weaving.

iv. 5000-4000 BCE: Cotton was cultivated. The cotton industry was well developed.Some methods of cotton spinning and fabrication continued to be practiced till the modern Industrialization of India. Hemp was also domesticated by the Indians. Hemp was used for a number of applications including making narcotics, fibre and oil. The farmers of the Indus Valley started to cultivate peas, sesame, and dates. Sugarcane was originated in tropical South Asia and Southeast Asia. Different species seems to originate in different locations with *S. barberi* originating in India and *S. edule* and *S. officinarum* coming from New Guinea.

v. 500 BCE: Irrigation was developed in the Indus Valley Civilization. The development of the Indus civilization grew as a result of this innovation, which led to more planned settlements making use of drainage.

4000 to 2000 years ago

i. In North Arcot, bone / stone tools were found.

ii. In Nevasa (Maharastra), copper and polished stone axes were used. First evidence of the presence of silk was found at this location.

iii. At Navdatoli on Narmada river (Nemar, Madhya Pradesh), sickles set with stone teeth were used for cutting crop stalks. Crops grown were linseed, lentil, urd (black gram), mung bean, khesari, and wheat.

iv. In Eastern India, rice, bananas, and sugarcane were cultivated.

2000-1500 years ago

i. Tank irrigation was developed and practiced widely.

ii. 3000 BCE: Sophisticated irrigation and water storage systems were developed by the Indus Valley civilization, including artificial reservoirs at Girnar.

iii. 2600 BCE: An early canal irrigation system from Circa.

iv. 2500 BCE: Archeological evidence of an animal-drawn plough in the Indus Valley civilization.

v. 2000 BCE: Agricultural activity included rice cultivation in the Kashmir and Harrappan regions.

vi. Greek and Romans had trade with South India; pepper, cloth, and sandal wood were imported by Romans.

vii. Chola King Karikala (190 AD) defeated Cheras and Pandyas, invaded Srilanka, captured 12000 men and used them as slaves to construct an embankment along the Cauvery, 160km along, to protect land from floods.

B. Vedic period-Post Maha Janapadas period (1500 BCE–200 CE)

i. Gupta (2004) finds it likely that summer monsoons may have been longer and may have contained moisture in excess than required for normal food production. This excessive moisture would have been added to the winter monsoon rainfall required for winter crops.

ii. In India, both wheat and barley are held to be Rabi (winter) crops and like other parts of the world—would have largely depended on winter monsoons before the irrigation became widespread. The growth of the *Kharif* crops would have probably suffered as a result of excessive moisture.

iii. Jute was first cultivated in India, where it was used to make ropes and cordage.

iv. Some animals thought by the Indians as being vital to their survival came to be worshiped.

v. Trees were also domesticated, worshiped, and venerated Pipal and Banyan in particular.

iv. Others came to be known for their medicinal uses and found mention in the holistic medical system Ayurveda.

vii. 1000–500 BCE: There are repeated references to iron. Cultivation of a wide range of cereals, vegetables and fruits is described. Animal husbandry was important part in Agriculture; Meat and milk products were part of their daily food. The soil was ploughed several times. Seeds were broadcasted. Cropping system was recommended. Cow dung provided the manure. Irrigation was practiced.

viii. 322–200 BCE: The Mauryan Empire categorized soils and made meteorological observations for agricultural use. Other facilities included construction and maintenance of dams, and horse-drawn chariots was also available.

ix. 300 BCE: The Greek diplomat Megasthenes, in his book Indika provides a secular eyewitness account of Indian agriculture.

C. Early Common Era – High middle ages (200–1200 CE)

i. Wide range of crops such as rice, sugarcane, millets, black pepper, various grains, coconuts, beans, cotton, plantain, tamarind, jackfruit, coconut, palm, areca and plantain trees and sandalwood were started to grow by the Tamil people.

ii. Systematic ploughing, manuring, weeding, irrigation and crop protection was practiced. Water storage systems were managed during this period.

iii. Kallanai (1st-2^{nd}century CE), a dam built on river Kaveri during this period, is considered as one of the oldest water-regulation structures in the world still in use.

iv. India started shipping spices to the Mediterranean with the spice trade involving native spices like cinnamon and black pepper gained momentum.

v. Roman trade with India followed as detailed by the archaeological record and the Periplus of the Erythraean Sea.

vi. Chinese sericulture attracted Indian sailors during the early centuries of the Common Era.

vii. 320-550 CE: The Guptas discovered crystallized sugar and the earliest literature showed that candied sugar come from India.

viii. 647 CE: Chinese people tried at least two missions to India for obtaining technology for sugar-refining, as per Chinese literature. Cereals such as wheat, rice and millets, and fruits were extensively cultivated. A 60daysshort duration variety and fragrant rice varieties are mentioned. Ginger, pumpkin, garlic, melons, onion, and mustard are also mentioned. Persian wheel was used in Thanesar (Haryana).

ix. Cholas era was a glorious phase in South Indian in the 10th century AD. New irrigation systems for agriculture were developed- chain tanks in Andhra in the 9th century; and 6.4km Kaveripak bund. Cholas had good communication with China, Myanmar, and Campodia. The tank supervision committee (Eri-variyam) looked after the maintenance of a village and regulated the water supply.

x. 875-1200 CE: Noboru Karashima's research of the agrarian society in South India during the Chola Empire reveals that during the Chola rule land was transferred and collective holding of land by a group of people slowly transferred to individual plots of land, each with their own irrigation system. These led to a decrease in areas of dry cultivation. The Cholas also had bureaucrats which oversaw the distribution of water, particularly the distribution of water by tank-and-channel networks to the drier areas.

D. Late middle ages – Early modern era (1200 - 1757 CE)

i. The construction of water works and water technology aspects in India is mentioned in Arabic and Persian works. Irrigation systems were developed due to the diffusion of Indian and Persian irrigation technologies which bought about economic growth and growth of material culture.

ii. Agricultural 'zones' were broadly classified into those producing rice, wheat or millets.

iii. Rice cultivation was dominated in Gujarat and wheat dominated north and central India.

iv. Many crops introduced to India during this period of extensive global discourse as mentioned in the Encyclopædia Britannica.

v. 1556-1605 CE: Land management was particularly strong during the regime of Akbar under whom scholar-bureaucrat Todarmal formulated and implemented elaborated methods for agricultural management on a rational basis.

vi. Indian crops—such as cotton, sugar, and citric fruits—spread throughout North Africa, Islamic Spain, and the Middle East.

E. Colonial British Era (1757–1947 CE)

i. Few Indian commercial crops such as cotton, indigo, opium, and rice- made it to the global market under the British Raj in India.

ii. The second half of the 19thcentury saw some increase in land under cultivation and agricultural production expanded at an average rate of about 1% per year by the later 19th century.

iii. Due to improved and extensive irrigation by canal networks, Punjab, Narmada valley, and Andhra Pradesh became centres of agriculture. The British regime in India developed irrigation works on the basis of scale required. Community effort and private investment also worked in market for irrigation development.

iv. Agricultural prices of some commodities rose to about three times between 1870 and 1920.

v. 1871: Government of India created Department of Revenue, Agriculture and Commerce which formed as base for initiation of Agriculture in India.

vi. 1880: Famine Commission Report was submitted which was base for inception of Agricultural Department.

vii. 1881: Separate Department of Agriculture at Centre for Famine relief operations was created.

viii. 1890: Dr. J.A. Voelcker appointed as a consulting chemist from Royal Agricultural Society (England) - Laid foundation for agricultural research in India.

ix. 1892 – 1903 - Appointment of Imperial Agricultural Chemist, Imperial Mycologist and Imperial Entomologist was started which was thebase for Beginning of inducting the scientist in Agriculture.

x. 1900: Lord Curzon passed the Land Foundation Act.

xi. 1904: Lord Curzon passed the Cooperative Societies Act

xii. 1901-05: To enhance agricultural education, Agricultural Colleges at Pune, Kanpur, Sabour, Nagpur, Coimbatore and Lyallpur (Now in Pakistan) were established.

xiii. 1905: Imperial Agricultural Research Institute (IARI) at Pusa (Bihar) was established. Lord Curzon, the then Viceroy of India with the generous donations and help from Henry Phipps of the USA, had set up the Imperial Agriculture Research Institute in 1905 at Pusa, a village in the Darabhanga district of Bihar. The name of the main building at Pusa was mentioned after its donor as the Phipps Laboratory. [PUSA stands for the donor of the Institute, Phipps of the USA].

xiv. 1926: The Government of India appointed a Royal Commission to examine the condition of agricultural and rural economy in India.

xv. 1929: Based on Royal Commission on Agriculture's recommendation (1928), Imperial Council of Agricultural Research (ICAR) was establishment as a Society under the Societies Registration Act, 1860to conduct comprehensive research on Agriculture. The Society was registered on July 16, 1929. [After Independence, the name of the society was changed to Indian Council of Agricultural Research (ICAR)].

xvi. 1936: There was a disastrous earthquake and Pusa suffered heavily. Therefter, the Government of India carefully considered and rebuilt the institute at New Delhi. The transfer to New Delhi was completed by October, 1936. In November, 1936, the Marquees of Linlithgo, the then Viceroy of India, opened this Institutes.

xvii. 1931-47: Indian Lac Cess Committee, Indian Central Tobacco Committee, Indian Central Oilseeds Committee were formed to improve research in various crops.

F. Republic of India (1947 CE onwards)

i. Special programs were undertaken to improve food and cash crops supply.

ii. The Grow More Food Campaign (1940s) and the Integrated Production Programme (1950s) focused on food and cash crops supply, respectively.

iii. 1957: All India CoordinatedMaize Improvement Projectwas initiated (First coordinatedproject) to exploit maize research (Specifically heterosis).

iv. Five-year plans of India oriented towards agricultural development soon followed.

v. 1963: Introduction of semi dwarf wheat varieties from CIMMYT, Mexico Formed basis for green revolution.

vi. 1966: Introduced semi-dwarf rice varieties TN1 & IR 8 from Taiwan and Philippines respectively is formed as base for green revolution. Land reclamation, land development, electrification, use of chemicals fertilizers in particular, mechanization, and development of agricultural 'package approach' of taking a set of actions instead of promoting single aspect followed under the supervision of government.

vii. The many 'production revolutions' initiated from 1960s onwards like Green Revolution, Yellow Revolution (oilseed: 1986-1990), Operation Flood (dairy: 1970-1996), and Blue Revolution (fishing: 1973-2002) etc.

viii. 1979: National Agricultural Research Project (NARP) was launched to strengthen the research capabilities of SAUs.

ix. Following the economic reforms of 1991, significant growth was registered in the agricultural sector, which was by now benefiting from the earlier reforms and the newer innovations of Agro-processing and Biotechnology.

x. 1998: National Agricultural Technology Project (NATP) was initiated to strengthen the research on location specific problems Contract farming which requires the farmers to produce crops for a company under contract and high value agricultural product increased.

I.2006: National Agricultural Innovative Project (NAIP) was launched for end to end approach for solving problems.

Probable age-old crop introductions

Abelmoschus esculentus (L.) Moench, Okra

The brand Abelmoschus is of Asiatic origin, however, the exact affiliated home of the able *A. esculentus* is acknowledged amid Africa and India. Plentiful affirmation suggests India to be one of its centers. Tindisha and Gandhamula, supports its age-old cultivation, because of earlier accession from Africa through a sea route.

Allium cepa L., Onion

The Central Asiatic arena is advised primary centermost of agent of onion admitting the abreast east Asiatic and Mediterranean regions are the accessory centers. The age-old annual of onion agriculture arise from Egypt. In India, its advertence is beginning in 6th period BC with Sanskrit names, Palanduh and Sukanda. Archaeo-botanical charcoal of *A. cepa* accept been recorded in India.

Allium sativum L., Garlic

Garlic is as well of Central Asian origin. The barter accord amid Mesopotamia, Egypt, and India during Mohenjo-Daro period, about 3000 BC, led to its accession into India. Later, the bulbs as berry were brought and grown. The Arabian traders and the Greeks (Alexander the 300 BC), 'Yavans', brought garlic's use in affable as an additive and into Ayurveda adjustment of medicine;

Amaranthus hypochondriacus L.; syn. A. leucocarpus

A built-in of the arctic Mexican highlands food came to India three millennia BC, as is adumbrated by its acceptable use and is broadly able as food crop in the Himalayan region. Tanduliya, acceptation 'reportedly a Land weed', Kataib and Chaulai; Another species, *A. cruentus*, appears to be an afterwards accession and *Amaranthus spinosus* L., the acicular amaranth, a built-in of Americas, is broadly broadcast as edger throughout south India and beyond.

Anacardium occidentale L., Cashew nut

A built-in of Brazil or Venezuela, cashew nut is about believed to accept been food into India in the 16th period AD (1560) by the Portuguese (Sauer, 1993). However, delineation of complete bulb with flowers and fruits at the Jambukeshvara temple in Tiruchirapalli, Tamil Nadu, congenital 2500 years back and sculptural delineation of cashew nut bake-apple at the Bharhut Stupa, anachronous ca. 200 BC, beforehand its age-old cultivation. It has Sanskrit names Beejara Sula or Bijara Sala, Shoephahara and Kajutaka.

Ananas comosus (L.) Merr., Pineapple

A built-in of Brazil, pineapple was food into India, apparently by the 5th period AD via Average East. A carve depicting its bake-apple has been begin on the vanamala of Vishnu in his Varaha avatara in the Udayagiri cavern temples in Madhya Pradesh (MP), anachronous 5th period AD. Reportedly originated its Sanskrit names where Bake-apple is acclimated alleviative purpose too.

Annona cherimolia Mill., Ample Annona, Custard Apple

Custard prime is built-in of the highlands of Colombia, Ecuador and Peru and was reportedly food into India in 1820. However, the bake-apple of A. cherimolia is apparent captivated by a sculpted goddess bulk on a bank of a Hoysala Dynasty temple in Karnataka, anachronous to 13th century.

Annona reticulata L., Bullock's Heart, Custard Apple

A built-in of abutting America, custard prime is aborigine in India to an admeasurement that some authors advised it a native. It has Sanskrit names Rama Sita, Luvunee, Ramphala and Krishnabeejam.

Capsicum annuum L., Chili Pepper

A built-in of South America, chili pepper is now able throughout India. It is about believed to accept been food into India by the Portuguese in 16th period AD. However, Chilies accept been developed and acclimated in India abundant earlier, as chilies are mentioned in Siva and Vamana Puranas, anachronous ca. 6th-8th period AD. In Sanskrit, it has been referred to as Marichiphalam, Katuvira, Katuirah and Rakta maricha.

Carthamus tinctorius L., Safflower

Genus *Carthamus* is broadcast from Spain and North Africa beyond the Average East to North India. Most affirmation suggests agent of C. tinctorius in an across apprenticed by the eastern Mediterranean and Persian Gulf. Age-old Sanskrit scriptures declared it as Kusumbha. Its flowers accept been acclimated in alertness of Ayurveda medicines.

Cicer arietinum L., Chickpea (Bengal gram)

It is a beating crop, earlier developed in Turkey about 7000 BC, and was afterwards food into India as earlier as 2000 BC. It was arise the attendance of chickpea (radiocarbon anachronous at 5450 BC) and provided affirmation of its agriculture in the Mediterranean basin during 3000-4000 BC. In age-old Sanskrit writings, it has been referred to as Chanaka.

Citrullus lanatus (Thunb.) Watermelon

It is a built-in of southern Africa. Archaeologically, colocynth seeds accept been anachronous to about 5800 BP and those of watermelon at about 4000 BP in Nile Valley. By 4th-7th period AD, it accomplished India with acclimatization of both agrarian *C. colocynthis* and *C. lanatus,* in the western plains of the subcontinent, authoritative it an accessory centermost of biogenetic diversity. Sushruta mentions it in his Sushruta Samhita (6th period BC) for agriculture forth the banks of the Indus River. In Sanskrit writings, it has been referred to as Tarambuja. Krishnabija, Latapanasa, Madhuraphala and Alpapramanaka; Seeds of watermelon accept been begin in the archaeological charcoal acknowledging its age-old cultivation.

Corchorus olitorius L., Tosa jute

The beyond biogenetic assortment aural the breed credibility to Africa as the primary centermost of its origin, with India or Indo-Myanmar arena getting the accessory centermost of diversity, across it has been able for centuries. From Africa, *C. olitorius* beforehand reached to India via Egypt and Syria in age-old times. Assay of jute bolt (not Tosa jute) at Harappa suggests use of jute in bolt accumulation in the subcontinent during the aiguille of urbanization in the Indus Valley. In Sanskrit, it has been referred to by names such as Patta and

Kakka Tapattraka.

Cucumis melo L., Muskmelon: Musk allotment is a built-in of Africa with populations extending to south-western Asia. Indian Subcontinent is the earlier home of surrounding variants and the accepted groups, apparently resulted from re-domestication of animals forms in earlier times. In Sanskrit, it has been referred to as Kalinga, Kharvuja and Madhuphala. The seeds of muskmelon accept been begin in archaeological charcoal acknowledging its age-old cultivation.

Cucurbita pepo L., Pumpkin: Pumpkin is a built-in of southern allotment of North America. But it is broadly able all over India as a backyard garden crop. It is mentioned in the Atharvaveda dating before 800 BC; in the Buddhist Jatakas and added texts as earlier as the 4th period AD: and as an additive of Ayurveda anesthetic in an Indian argument of the 9th period and in medieval text. It has Sanskrit names, Kurkaru and Kushmanda.

Eleusine coracana (L.) Gaertn., Feel millet (Ragi): The centermost of conduct of feel millet is controversial based on the anthologized diversity, Vavilov (1935) appropriate India to be one of its homelands. Whereas, other reported that based on accident of its apparent ancestor, the agrarian tetraploid *Elusine africana*, which crosses advisedly with able feel millet; classification and accent appropriate African origin. Almanac of carbonized seeds of able feel millet associated with Iron Age in Zimbabwe (Inyanga) corroborated it.

Gossypium barbadense L., Tetraploid cotton: A built-in of South America, tetraploid affection was into India via Pacific to Indus civilization apparent in agriculture and address of cotton. The Sanskrit name is Maghani and its ability to accept accomplishment before 1000 AD, and came into cultivation.

Guizotia abyssinica (L.f.) Cass., Niger: Niger originated in the Ethiopian highlands as an oilseed crop. It is believed to accept been brought to India by the Ethiopian immigrants, apparently in the 3rd millennium BC forth with crops such as feel millet and its agrarian relatives. It has been referred to as Rama Tilah. There are no archaeo-logical finds of niger in age-old Sanskrit writing,

Helianthus annuus L., Sunflower: A built-in of Central and South Americas, sunflower is able all over India from actual earlier times as an accessory and apparently as well as an oilseed crop. It is mentioned in the Charaka's Charaka Sahmita anachronous no afterwards than the 4th period AD. Its Sanskrit names are Suriya-mukhi, Adityabhakta and Suryamukhi. Clear delineation of sunflower begins in Rani Gumpha cave, Udaigiri in Odisha anachronous to the 2nd period BC. Carvings of sunflower from the up temples of Hoysala Empire (10th to 14th century) in Karnataka was as well evidenced.

Hibiscus cannabinus L., Kenaf or Deccan hemp: Kenaf is believed to accept originated in age-old Africa, occurring as earlier as 4000 BC. It beforehand reached to India at an earlier period via Persia. It origin to south Asia, so plentiful so that several workers accept advised it native. It has been referred in Sanskrit conceptual of age-old India of 7th period with name, *balamulika.*

Hordeum vulgare L., Barley

Barley was one of the earlier grains peaceful in plentiful Crescent (Crescent appearance clammy ground consisting of Western Asia, the Nile Basin and delta). The management of agrarian barley (*Hordeum vulgare* ssp. *spontaneum*) ranges from North Africa and Crete in the west to Tibet in the east. It was unfamiliar to North India in age-old times and was apparently the semicircle food item in earlier days. Barley is mentioned in earlier Sanskrit with names such as Yava, Sithashuka, Atiyava and Akshata.

Ipomoea batatas (L.) Lam., Sweet potato: This built-in of South America is able all over India as poor-man's food. Sweet potato was mentioned in Sanskrit literatures as "Valli", Pindalah and Raktaluh, which are agnate to the names acclimated in northwestern part of South America.

Lagenaria siceraria (Molina) Stands Bottle gourd: Bottle gourd is one of the oldest plants, which originated from agrarian community in southern Africa, and ability to accept troubled from there to Asia by amphibian reference during summer shower or with animal migration. A biogenetic appraisal found on archeological samples appropriate that it may accept calm beforehand than food crops and livestock. Based on the appraisal of archeological samples, advisers announce two different domestications: one 8-9,000 years ago in Asia. In Sanskrit, it is referred to as Alabu and Kutukumbi (Pullaiah, 2002) acknowledging its age-old cultivation.

Lathyrus sativus L., Grass pea (Khesari): It is accepted to accept originated in Southwest and Central Asia or the Balkan Peninsula in the earlier Neolithic period. Grass pea may accept the earlier crop calm in Europe about 6000 BCE. In Sanskrit literature, it has been referred to as Thriputa, Kalaaya, Khandikaa, Lanka, Sandika and Dirghamoola. Seeds of *L. sativus* has been accepted and begins in archaeological charcoal at Malhar in UP, Balu, Kunal in Haryana, and average Gangetic Plains incorporation its age-old cultivation.

Lawsonia inermis L., Hena: Hena is built-in to Arctic Africa and western and southern Asia. In age-old Egypt, about 6,000 years ago, it was acclimated as a remedial hair dye. Whether it was brought into use at the time or before or afterwards in India is not known, but the use of Mehendi and turmeric has been declared in Vedas. The actuality and use of Henna for the endure 5,000 years ago can as well be begin in the books of Charaka and Sushruta and additional texts, and in the paintings and sculptures. In Sanskrit, it has been referred to as Mendika, Ragangi and Raktgarbha. It is getting acclimated as derma and hair dye in India back continued time as is apparent from the paintings of acclaimed Ajanta-Ellora caves.

Lense culinaris Medik., Lentil: Lentil originated in the plentiful Crescent and again beforehand to the Indo-Gangetic Plain. Historians appropriate its accession with Sanskrit speaking people, the Aryans, based on non-tenable approach of Indo-Aryan migration. In Sanskrit, it is referred to as Masura, Renuka, Mangalya. Seeds of lentil has been accepted in archaeological charcoal at Malhar in UP, Balu, Kunal in Haryana, Average Gangetic Plains acknowledging its age-old attainment and cultivation.

Linum usitatissimum L., Beat or Linseed: Wild beat grows from Arctic Africa to India, admitting it was earlier calm in the plentiful Crescent region. The age-old affirmation of agrarian beat use as a secure crop comes from Republic of Georgia, across spun dyed and askance agrarian beat fibers begin in Dzudzuana cave, anachronous 30,000 years ago in High Paleolithic period. In India, there is affirmation for its agriculture 5,000 years ago (3000 BCE). It has several Sanskrit names such as Uma, Kshuma, Atasi, Neel Pushpin, and Archaeo-botanical charcoal of beat or linseed berry accept begin at Harappa and abroad in North India.

Medicago sativa L., Lucerne, Alfa-alfa: Originated in South-Central Asia, Lucerne is frequently believed to be unfamiliar crop into India in 1900. However, it is advised Ladakh as one of the centers of its origin. Archaeo-botanical assessment of seeds at Kunal in Haryana confirms its age-old cultivation/introduction in the country.

Morus alba L., White mulberry: White mulberry is built-in to arctic China, Japan, and India. However, some accept it continued from Northwestern India into Asia Accessory and Persia. This timberline is able in all the originality of inventions of China for 4,000 years for appearance of silk-worms and authoritative cottony fabric. It was food into India from China during the age-old times, and is now developed broadly in the subcontinent. The age-old treatise, *Arthashastra* mentions Assam's growing cottony trade. Bush agriculture of white mulberry in Bengal for agriculture silk-worms is able known. In Sanskrit, it is alleged Tula.

Nicotiana tabacum L., Tobacco: A built-in of the Americas, as per Materia Medica, tobacco is an agricultural crop as food into India in 1605 AD. Others accept that it began in the New Prime and was agitated to India afterwards its accession into Europe. However, the use of the water-cooled smoker device, 'Hooka', is apparent in pre-Columbian art of India, forth with references to tobacco in acceptable medical practices. Its Sanskrit name is Tamakhu.

Archaeologically, a temple in Himachal Pradesh anachronous to 1422–1424 AD depicts the use of the Hooka.

Pachyrhizus erosus (L.) Urb., Yam bean: A built-in to the Americas, yam bean was food into India at atomic before 1000 AD and apparently abundant earlier. In Sanskrit, it is alleged Sankhálu. It is able throughout India for its ample comestible root, but does not activity in agrarian acceptance its earlier introduction.

Panicum miliaceum L., Proso millet: Both, the agrarian anterior and the across of conduct of proso millet, are unknown. It earlier appeared apart as a crop in China about 7,000 years ago. It is an earlier accession into India, and is in large quantities able in the country. In Sanskrit, it is called, Cheenaka, Kakakangu and Kangu.

Paspalum scrobiculatum var. polystachyum; syn. P. scrobiculatum L., Kodo millet: It originated in adjoining Africa and apparently travelled to Indian subcontinent beyond the Indian Ocean may be by accustomed activity like wind. It was calm in India 3,000 years ago and is still getting calm in Deccan Plateau, across it has become a food crop.

Pennisetum glaucum (L.) R. Br., Fair millet: Based on the attendance of agrarian relatives, southern bend of Sahara has been advised across of conduct of fair millet. It was food into India actual early, apparently by sea-route, as it is not begin in Nile Basin and Abreast East for achievability of a Land transfer. In Sanskrit, it is called, iksupatra, vata. Well-founded archeological charcoal of fair millet (1000-2200 BC) accept been acquired from several sites in Harapan culture, Gangetic Plains, and Deccan Peninsula, such as Rangpur in Gujarat, Narhan in UP and Hallur in Karnataka by 1500 BC.

Phaseolus vulgaris L., French or Kidney bean (Rajma): *Phaseolus vulgaris* has American origin. In India, it is able all over, but not accepted in agrarian. Archaeologically, and based on Sanskrit name S*imbi*, it was present in the southwest Asia and India all the way back to Sumerian times. Archaeological charcoals begin from pre-Prabhas and Prabhas cultures at Prabhas Patan in Gujarat dating 1800 BC to 600 AD; from post-Harappa, Chalcolithic website Inamgaon and additional places.

Phoenix dactylifera L., [P. sylvestris (L.) Roxb., Indian, Date palm]: Exact agent of date palm approach is unknown. It apparently originated about Iraq and has been able back age-old times (4000 BCE) in Mesopotamia to Egypt. It is originated and broadly able in India forth with surrounding breed Phoenix sylvestris from age-old time. Affirmation of its agriculture is beginning in the Indus Valley, including the Harappa period 2600 to 1900 BC. Rock charcoal of date accept been begin from Rohira in Punjab, and Balu in Haryana.

Pisum arvense L., Land pea: Native to the Mediterranean arena and North Africa, Land pea is broadly developed for forage. It was domesticated about 7,000 years ago, as adumbrated by the seeds begin in archaeological sites in Turkey. In Sanskrit, it is referred to as Sitla. The seeds of Land pea accept been begin at Harappa and at several Neolithic sites in Gangetic Plains.

Pisum sativum L., Pea: The agrarian pea is belted to the Mediterranean basin and the Abreast East, and was apparently calm there. The age-old archaeological study finds of pea anachronous from the backward Neolithic era of accepted in Greece, Syria, Turkey, and Jordan. In the additional bisected of the 2nd millennium BC, it appeared in the Gangetic basin and southern India. In Sanskrit, it has been referred to as Renuka, Satila, Triputa, Ananu and Hrenu. Archaeological charcoal accept been begin in Harappa and North West India dating back to 2250-1750 BC.

Psidium guajava L., Guava: Guava is a built-in of abutting America. The website of Caral in the Supe Basin of Peru yielded charcoal of guava, anachronous amid 2627 and 2020 Cal BC. It is aborigine and able all over India, and has several Sanskrit names such as Perela, Mansala, Amruta-phalam and Perukah. It finds acknowledgment in Charaka Samhita, advertence that it was food not afterwards than the 4th period or appreciably earlier, adverse the accepted acceptance for its accession by the Portuguese in 17th Century.

Ricinus communis L., Castor: Domesticated in West-Central Africa, brush beforehand to Egypt (4000 BC) and again accomplished India (2000 BC). In Sanskrit, it is referred to as Eranda, Gandharva hasta and by few added names in Bhava Prakash Nighantu. The age-old archaeobotanical almanac of brush in India was begin at the backward Harappa site, Hulas anachronous amid 1800-1300 BC. Back then, it has been recorded both in Indus basin and Gangetic Plains. The use of brush oil in India for lighting lamps and in surrounding anesthetic as a laxative, purgative, and abrasive has been accurate back to 2000 BC.

Setaria italic (L.) P. Beauv., Foxtail millet: Foxtail millet is a built-in of China with history of agriculture dating back to the Neolithic Era over 4,000 year ago. It abashed from China to south and westward to ability India. In South India, it is a basal diet for the humans from the Sangam period. In Sanskrit, it has been referred to as Bhvajja, Priyagguka, Rajika, etc. acknowledging its age-old cultivation. It occurred in Harappa levels (2500-2200 BC) at Shikarpur (Kutch). Archaeobotanical charcoal accept as well been begin from high and average Gangetic Plains.

Sorghum bicolor (L.) Moench. Sorghum: Most studies accede with African agent of sorghum with conduct accession from Southwest Ethiopia to Lake Chad. It accomplished India in pre-historical time. It is broadly developed in

India. In Sanskrit writings, it is alleged Yavanal. There accept been suggestions for reimport of cultivars developed in India.

Tagetes erecta L., *T. patula* L. (dwarf), Marigold: A built-in of South-central Mexico, some accept marigold originated in Africa. It accustomed in India while Sanskrit was the animate language, and is referred to as Zanduga, Sandu, Sthulapushpa and Ganduga. There is plentiful confirmation to beforehand that it was developed in India in old-age continued time and was associated with Hindu religious ceremonies for decoration and coloring, about a agnate use as in Mexico.

Tamarindus indica L., Tamarind: Tamarind is of West African origin, but it has beforehand to India at an earlier date. There are abounding references to the tamarind timberline in Hindu mythology. In Sanskrit, it has been referred to as Amlika, Sucukrika, Tintrini and Yamadutika. Charcoal charcoal of tamarind timberline accept been articular from Narhan website of the average Gangetic Plains dating back to 1300 BC.

Trifolium alexandrinum L., Berseem: Berseem is a built-in of Egypt. As per accepted belief, it was food into North India in the earlier 19th century. But the age-old cultural and barter contacts amid Egypt and India, its advertence in Sanskrit autograph as Tripatra, and award of three seeds in archaeobotanical charcoal from complete Harappa phases abate this anticipation and announce its age-old introduction/cultivation.

Trigonella foenum-graecum L., Fenugreek: A built-in to eastern Mediterranean region, fenugreek was brought into agriculture in the Abreast East as adumbrated by the broiled fenugreek seeds recovered from Tell Halal in Iraq dating back to 4000 BC. It is begin agrarian in Kashmir, Punjab and the high Gangetic Plains suggesting its acclimatization to India. In Sanskrit, it is referred to as Medhika, Methika, Chandrika and several added names, acknowledging its age-old cultivation.

Triticum - *T. aestivum* L. ssp. *aestivum*, accepted Hexaploid Wheat: Common as was earlier calm in Western Asia (Fertile Crescent—upper alcove of the Tigris-Euphrates arising basin) annular 11,700 years BCE. Its agriculture began to beforehand afterwards about 8000 BCE (9,000 years back) and accomplished India, across it acquired further. For this reason, the north-western end of Indian Subcontinent is admired as the accessory centermost of origin, arch to agent of T. sphaerococcum (dwarf wheat). Blasting at Mehrgarh arise species, T. monococcum, T. dicoccum and T. durum or T. aestivum, suggesting that Baluchistan was a actual earlier centermost of aureate cultivation. In Sanskrit, it has been referred to as Godhuma, Sumana, Dhuma, Kanika, etc. Carbonized aureate grains apparent at Mohenjo-Daro are anachronous to 1755 BC.

Vigna unguiculata (L.) Walp., Cowpea: Cowpea is earlier to West-Central Africa. Its adept agriculture in India led to an acceptance for its Asian origin. In Sanskrit, it has been referred to as Harimanth, acknowledging its attendance in India before the Christian era. Archaeobotanical charcoal accept been begin at Daimabad in Maharashtra ca.1700-1500 BC, Hulas in high Gangetic Plains 2200-1500 BC and in Khairadih in average Gangetic Plains (Saraswat, 2005).

Vitis vinifera L., Grapes: A built-in of Eurasia, grapes is anticipation to accept been calm earlier in Southwestern Asia during the Neolithic period. In Asia, the age-old affirmation of agriculture of grapes is from Iran and Baluchistan, anachronous before 2000 BC, wherefrom agriculture boring beforehand to India and China with the alpha of Christian era. It is not mentioned in Vedic abstract as Draksha.

Zea mays L., Maize: Maize is about believed to accept been food into India by the Portuguese during the 17th century. However, the assay of Sikkim Primitive authenticate that maize was present in India in age-old Era and was able abundant before the about-face of Christian era. In Sanskrit, it is accepted as Sasyam, Stambakari, Sasyavisesha, besides Yavanala. References to maize fields are begin in the inscriptions of Assam (kostha makkhi yana). Sculptures of maize-ears were beginning in the Hoysala Empire temples in Karnataka between11th-13th centuries.

Crops of Indian origin

Besides, there are assertive crops which were built-in of our country and in the beforehand of crop boating which accept been beforehand beyond the globe. Some of such crops are listed below.

Oryza sp., Rice: India is an important centre of rice cultivation. The rice is able on the better areas in India. Historians accept that while the indica array of rice was earlier calm in the across accoutrement the foothills of the Eastern Himalayas (i.e. north-eastern India), accession through Burma, Thailand, Laos, Vietnam and Southern China, the japonica array was calm from agrarian rice in southern China which was food to India. Abiding agrarian rice still abound in Assam and Nepal. It seems to accept appeared about 1400 BC in southern India afterwards its conduct in the arctic plains. It again beforehand reached to all the abundant alluvial plains watered by rivers. Some says that the chat rice is acquired from the Tamil chat arisi. Rice is earlier mentioned in the Yajur Veda (c. 1500-800 BC) and again is frequently referred to in Sanskrit texts. In India there is an adage that grains of rice should be like two brothers, abutting but not ashore together. Rice is generally anon associated with affluence and fertility; appropriately there is the custom of throwing rice at newlyweds. In India, rice is consistently the earlier food offered to the babies if they alpha bistro debris or to bedmate by his new bride, to ensure they will accept children.

Aegle marmelos, Beal: Aegle marmelos, frequently accepted as bael, as well Bengal quince, aureate apple, Japanese absinthian orange, rock prime or copse apple, is a breed of timberline built-in to India, Nepal, the Andaman and Nicobar Islands and Myanmar. It is present in Sri Lanka, Thailand and Malaysia as an original species. The timberline is advised to be primeic by Hindus. Its fruits are acclimated in acceptable anesthetic and as a food throughout its range. The accepted name "wood apple" may as well accredit to *Limonia acidissima*.

Piper nigrum, pepper: Black pepper (*Piper nigrum*) is a beginning backcountry in the ancestors Piperaceae, able for its bake-apple which is usually broiled and acclimated as a aroma and seasoning, accepted as a peppercorn. If beginning and absolutely mature, it is about 5 millimeters (0.20 in) in bore and aphotic red, and contains an individual berry like all drupes. Peppercorns and the arena pepper acquired from them may be declared artlessly as pepper, or added absolutely as pepper (cooked and broiled blooming fruit), blooming pepper (dried blooming fruit), and white pepper (ripe bake-apple seeds). Black pepper is built-in to south India and is abundantly able there and abroad in abutting regions. Currently, Vietnam is the world's better ambassador and exporter of pepper, bearing 34% of the world's *Piper nigrum* crop as of 2013. Pepper is built-in to South Asia and Southeast Asia and has been accepted to Indian affable back at atomic 2000 BCE. The absent age-old anchorage city-limits of Muziris in Kerala, acclaimed for exporting pepper and assorted added spices, gets mentioned in a bulk of classical actual sources. The bequest of this barter charcoal in some Western acknowledged systems that admit the appellation peppercorn hire as a badge transaction for something that is, essentially, getting given.

The age-old history of pepper is generally interlinked with (and abashed with) that of continued pepper, the broiled bake-apple of carefully accompanying *Piper longum*. The Romans knew of both and generally referred to either as just "piper". In fact, it was not until the assay of the New Prime and of chili peppers that the acceptance of continued pepper absolutely declined. Chili peppers, some of which, if dried, are agnate in appearance and aftertaste to continued pepper—were easier to abound in an array of locations added acceptable to Europe.

Before the 16th century, pepper was getting developed in Java, Sunda, Sumatra, Madagascar, Malaysia, and everywhere in Southeast Asia. These areas traded mainly with China, or acclimatize the pepper locally. Ports in the Malabar across as well served as a stop-off point for abundant of the barter in added spices from further east in the Indian Ocean. Afterward the British administration in India, around all of the atramentous pepper begins in Europe, the Average East, and North Africa was traded from Malabar region.

Solanum melongena, Eggplant: Eggplant (*Solanum melongena*), or brinjal or aubergine, is a breed of nightshade, developed for its comestible fruit. Eggplant is the accepted name in North America, Australia and New Zealand, but British English uses the French chat aborigine. The bulb breed originated in cultivation. It has been able in southern and eastern Asia back prehistory. The earlier accepted accounting almanac of the bulb is begin in Qimin Yaoshu, an age-old Chinese agricultural argument completed in 544.[7] The abundant Arabic and North African names for it, forth with the abridgement of the age-old Greek and Roman names, announce it was food throughout the Mediterranean across by the Arabs in the earlier Average Ages. The aubergine is accepted in England until the 16th century.

Gossypium arboretum, Timberline cotton: *Gossypium arboreum*, frequently alleged timberline cotton, is a breed of affection built-in to India, Pakistan and added abutting and subtropical regions of the Old World. There is affirmation of its agriculture as continued ago as the Harappan civilization of the Indus Basin for the accumulation of affection textiles. This breed of affection was as well food into East Africa and was developed by the Meroe civilization in Nubia. The brier was included in Linnaeus's Breed Plantarum arise in 1753. The holotype was as well supplied by him, which is now in the Linnean Herbarium in the Swedish Museum of Accustomed History. It is a sister breed of Gossypium herbaceum.

Artocarpus heterophyllus, Jackfruit: The jackfruit (*Artocarpus heterophyllus*), is as well accepted as jack tree, fenne, jakfruit, or sometimes artlessly jack or jak. The jackfruit timberline is a broadly able and accepted food account throughout the abutting regions of the world. Jackfruit is the civic bake-apple of Bangladesh. The jackfruit has played a cogent role in Indian agriculture for centuries. Archeological allegation in India accepts arise that jackfruit was able in India 3000 to 6000 years ago. It has as well been broadly able in Southeast Asia. Stingless bees such as *Tetragonula iridipennis* are jackfruit pollinators, and as such, play an important role in jackfruit cultivation. Outside of its countries of origin, beginning jackfruit can begin at food markets throughout Southeast Asia. It is as well abundantly able in the Brazilian littoral region, across it is awash in surrounding markets. It is accessible canned in bathetic syrup, or frozen, already able and cut. Broiled jackfruit chips are produced by assorted manufacturers. In arctic Australia, absolutely in Darwin, jackfruit can begin in alfresco after-effects markets during the dry season. Outside of countries across it is grown, jackfruit can be acquired year-round, both canned and dried. Jackfruit industries are accustomed in Sri Lanka and Vietnam, across the bake-apple is candy into articles such as flour, noodles, papad, and ice cream. It is as well canned and awash as a vegetable for export. The copse of the jackfruit timberline is important in Sri Lanka and is exported to Europe; it is termite-proof and is to teak for architecture furniture.

Cajuns cajan, Pigeon pea: The pigeon pea (*Cajanus cajan*) is a abiding legume from the ancestors Fabaceae. Back its conduct in India at atomic 3,500 years ago, its seeds accept become a accepted food atom in Asia, Africa, and Latin America. It is captivated on an ample calibration mainly in south Asia and is an anterior of protein for the population of that subcontinent. The agriculture of the pigeon pea goes back at atomic 3,500 years. The centre of agent is apparently peninsular India, across the abutting agrarian ancestors (*Cajanus cajanifolia*) activity in abutting deciduous woodlands. Archaeological finds of pigeon pea dating to about 3400 years ago (14th period BC) accept been begin at Neolithic sites in Karnataka (Sanganakallu) and its bound areas (Tuljapur Garhi in Maharashtra and Gopalpur in Orissa) and as well the south Indian states such as Kerala, across it is alleged Tomara Payaru. India from catholic to East Africa and West Africa, it is spread over. There, it was earlier encountered by Europeans, so it acquired the name Congo Pea. By bureau of the bondservant trade, it came to the American continent, apparently in the 17th century.

Curcuma longa, Turmeric: Turmeric (*Curcuma longa*) is a rhizomatous blooming abiding bulb of the gonger family, Zingiberaceae. It is built-in to Southeast Asia. Plants are aggregate annually for their rhizomes and broadcast from some of those rhizomes in the afterward season. Turmeric has been acclimated in Asia for bags of years and is an allotment of Ayurveda, Siddha medicine, Unani, and acceptable Chinese medicine. It was earlier acclimated as a dye, and again afterwards for its alleviative properties.

Ziziphus mauritiana, Indian plum: *Ziziphus mauritiana*, as well accepted as Chinese date, ber, Chinee apple, jujube, Indian plum, Regi pandu, Indian jujube, is a abutting bake-apple timberline breed acceptance to the ancestors Rhamnaceae. The breed is believed to accept originated in Indo-Malaysian arena of South-East Asia. It is now broadly aborigine throughout the Old Prime tropics from Southern Africa through the Average East to the Indian Subcontinent and China, Indomalaya, and into Australasia and the Pacific Islands. It can anatomy close stands and become invasive in some areas, including Fiji and Australia and has become an austere ecology edger in Arctic Australia. It is a fast-growing timberline with an average lifespan, that can bound ability up to 10–40 ft. (3 to 12 m) tall.

Conclusion

The assessment is to acclimate the appraisals of agricultural earlier and the attainment activity entry to assessment determinants of agricultural succession over the year 1950-51 through 2005-06. The incorporation indicates that there is to admission both net sown across and gross sown area. The incorporation as able highlights that discarded 39 percent of net sown across is rainfed area. Afterwards evaluating the changes in agrarian structure recognizes use adjustment and beforehand trend of agriculture, this point out some basal reasons. These are: agricultural workforce uncomfortable from cultivators to agricultural labours, the bulk of uneconomic abetment has an accession trend, across below food crops embarrassed arise non food crops, and rural food crops across below cereals has been animate arise non cereals, beforehand trend of accumulated agriculture as able as all sub across of agriculture except forestry is courageous collapsed trend during post-WTO period. It is as able empiric in this incorporation that succession and crop alternation below for about crop during procession alleviate while, across alternation added in the period. This additional information indicates that alternation in across became acquiescent bureau for accession instability. The atomization appraisal indicates that ascendance expertise per hectare is the complete anterior of agricultural beforehand for a lot of the crops and crop groups. Disaggregating of advertence period in four sub-periods shows that accession of agricultural land was the basal anterior of agricultural beforehand during the period before 1965-66 afterwards that the accession of added land across below agricultural accession has below over time and admission in affluence became the basal anterior of beforehand in agricultural production. The account of accumulated agricultural accession activity with both models indicates that land absolutely afflicted the agricultural adeptness beforehand during 1950-51 to 1964-65 and afterwards that land became beneath convincing and now labour and basal inputs are absolutely affecting the agricultural proficiency growth. Thus, the aftereffect to accumulate agricultural attainment activity verifies the after-effects of analysis.

Some pertinent short questions

1. What are the indices to analyze age-old agriculture?
2. What are the probable age-old crop introductions?
3. Write down five crops of Indian Origin;
4. Accustomed history of *Allium cepa* L., Onion;
5. Accustomed history of *Ziziphus mauritiana*, Indian plum;

9

Crop Production Through Indigenous Traditional Knowledge (ITKs)

Introduction

Research accomplishments on mentioned findings showed that the quality of community voice like access is familiarized for agricultural revamp index that adapted the charge to findings at the lower like as the antecedence to analyze a number of deficiencies of complete analysis works in a specific area. Firstly, the accomplishments arise to be alloyed and diverse. Secondly, with the agricultural achievement of a span afflict decidedly by its natural, institutional and technological characteristics (which alter broadly an allocation of states), absolute analysis absorption abundantly at the accompaniment like and the change of basal up and micro-level planning emphasizing the issue. Thirdly, there is an absence of studies on agricultural achievement at the disaggregate like in the accompaniment excepting a few that does put up with are bedfast to the community set up and are based on an actual leap amount of variables accouterment periods up to the original since the year 2000. Besides a lot of changes in agriculture pattern, irrigation coverage, agriculture intensity, fertilizer consumption, boilerplate measurement of operational backing and allotment of agricultural workers accept taken abode in agriculture during the contemporary years. These changes knowledge accept afflict the agricultural achievement of districts abnormally and causeless to say they could not accept been grasped in the absolute studies. There is appropriately a charge for advantageous these analysis gaps and adorning the absolute literature. The present findings are an apprehensive advance in the management and administration to apperceive of the ITKs attainable and its role in alteration over agricultural production. In adjustment to access the acknowledgment of the farmers apropos the ITKs uses in agriculture, the analysis agenda developed with a proforma that accomplished in advised and basic villages in altered agro-ecological regions through the activity centers. Farmers' social contact and acceptance of ITKs practices were attempted in the findings and recorded for an all-important symbol the abstracts to quantify the ITKs role in present-day agriculture.

Documentation of ITKs

The ITKs were accurate beneath afterward specific categories

- Agronomic Measures
- Tillage
- Bunding and Terracing (Mechanical and Vegetative barrier)
- Acreage Configuration
- Soil Amendment and Mulching
- Erosion Superiority and Runoff Diversion Structures
- Water Harvesting, Seepage Ascendancy and Arena Baptize Recharge

Possible stakeholders of the ITKs

The stakeholders who can be ally for announcement the ITKs in agriculture are as follows:

1. Farmers
2. NGOs
3. Government Addendum Agencies
4. Analysis Institutes / Scientists
5. Administrators
6. Action makers / people's representatives

Categories of ITKs

In familiar with the ITKs those are based on three categories viz., (a) cultural practices, (b) mechanical practices and (c) use of botanicals. Some adequate and forward-thinking knowledge in ambience with the all categories accepting application for acceptance in integrated crop management and best management practices and strategies to accept in principle and those have been advised and presented in the afterward.

Cultural practices: Generally cultural practices enhance the "belowground biodiversity" which accordingly accord to "aboveground biodiversity" and accomplish the abode added assorted for sustenance of accustomed enemies. The cultural practices (field sanitation; able berry and array selection; able seedbed preparation; burying date; row spacing; agriculture rate; fertilization; baptize management; crop rotation; burying of allurement crops and barrier rows; accompaniment planting; and intercropping) accord to prevent, suppress, or eradicate annoyance accession by abolition the accustomed accord amid the annoyance and the host ball and appropriately accomplish the annoyance beneath adequate to survive, grow, or reproduce. A lot of these practices are able-bodied experimented and accomplished by the farmers. However, some strategies to abound arena awning crop and windbreaks are adapted to abate dust because dust can baffle with accustomed enemies and may be able to cause outbreaks of pests such as spider mites. Similarly there is a charge to abstain balance fertilization and irrigation, which can could because phloem-feeding pests such as aphids to carbon added rapidly than accustomed enemies.

Mechanical practices: It includes able acreage preparation; hoeing; weeding, bagging of fruits; baits and traps; row covers; mulching; handpicking; and pruning, etc. However, with the acute demands for institution yields to abutment the growing citizenry an about-face from the adequate agriculture adjustment to accelerated agriculture took abode breadth appliance of constructed pesticide has become a broadly accustomed adjustment in annoyance control. Extensive use of top acquiescent varieties and agrochemicals like fertilizers and pesticides accompanying with the assorted impacts of altitude change has brought surprising changes in the crop.

Use of botanicals: The conveyance of pesticide application as an apparatus of original resort has now drifted the accepted best management practices and strategies from its ecological roots. Therefore, development of new knowledge systems with accent on eco-friendly approaches and new agricultural management strategies are acquainted with capital to anticipate new techniques or the attainment of the complete ones. It has already been accustomed that the sustain knowledge in crop cultivation and protection can discarded be accomplished by developing technologies that are based on locally attainable inputs that are calmly adequate and amount effective. The forward-thinking account of farmers in analytic technological problems by bounded knowledge managements are therefore, accepting accent in contempt years. Innovations based on adequate knowledge accept as well been emphasized by the National Information Foundation (NIF) and Department of Science & Technology (DST), Government of India. The World Summit on Sustaianable Development (WSSD) captivated in Johanesburg of South Africa during 2002, is has been acerb advocated the use of enclosed technological knowledge in crop husbandry amalgamation (Anon, 2002).

ITKs for crop cultivation

Some relevant ITKs popular are added findings as healthy as adaption, analysis and development of new projects. In appearance to findings in deep hole scientifically that could change these ITKs to Modern Technological Knowledge (MTK) befitting identities of prevailing ITKs. All the advancing projects on knowledge attention and management should focus on the applicable and adapted ITKs apropos to soil and water for adequate development and broadcasting of the surrounded technology. As an initiative, the ITKs for added agro sensitive regions may be correct and true afterwards and visual at delimited level. Exposure appointment and farmer-to-farmer alternation on the accountable may be encouraged for bigger adoption. A majority of farmers in a lot of developing countries are small-scale farmers and their knowledge systems accept anytime been recorded systematically appropriately they are not calmly attainable to agricultural researchers, addendum workers, and development practitioners. However, a typical efforts were fabricated in the contempt accomplished in affidavit of the ITKs in publications like "Honey Bee" (a journal); "Inventory of Indigenous Technological Knowledge in Agriculture" with two supplements (an accumulation of the mission mode activity of National Agricultural Technology Project on 'Collection, confirmation and validation of indigenous technological knowledge appeared by ICAR, New Delhi during 2003) and "Explored indigenous technological knowledge in displeasure management" (an advertisement of NCIPM, New Delhi, 2005). A lot of the accurate ITKs charged validation before they could be recommended to the agriculture community.

Preservation of ecological knowledge

Recording and documentation of traditional knowledge requires close participatory research with communities, as they help in identifying and preserving traditional knowledge in various ways. For example, there are a few trees and plants such as *tulsi* (*Ocimun religiosum*), which are considered sacred and worshipped by the people. The reason for this may be that such socially-valued trees must be of great use. As a result, they have been preserved in the name of religion.

Critically and scientifically validate the ecological knowledge by conducting research at grassroots levels and then start analyzing this knowledge for formulating strategies for better land utilization. A broad-based study should also be conducted to find out how traditional can be applicable worldwide. Development will automatically begin when scientific traditional practices are adopted. In many places, it had helped in setting up rainwater harvesting structures, starting from the Himalaya to Nagaland. A major change was observed due to these structures. The agricultural practices of the people improved without any assistance. Water acted as an agent in bringing about positive changes, leading to the development of a new agricultural system.

Needs for documentation

Many ITKs on in-situ in agriculture are not adopted everywhere throughout the country because of constraints in acceptance and blindness of the cap knowledge of such practices. The present confirmation activity has audible address on the approaching advance of activity in framing new projects. This concise affidavit activity may advance to the afterward approaching activities as follows:

1. Similar exercise can be undertaken to certificate the ITKs from all the agro ecological regions of the country.
2. The relevant ITKs may be activated for their adequacy and acceptance in added agro-ecological regions as a broadcasting strategy.
3. The accurate ITKs may be published/ translated in all delimited languages for the account of the agriculture communities.
4. Validation of the ITKs is an analytic step to authorize and quantify the restricted knowledge of these practices. Adequate modifications of the adequate practices through on-farm analysis would advise in developing adapted and adequate technologies for altered surrounded environments.
5. The aftereffect of attention measures on knowledge losses can be advised in detail through analysis and use of dispatch model.
6. As an action amount the delimited ITKs should be in inherited in the knowledge attention programme.
7. Hence, the activity of documentation, clarification and advance of ITKs should structure of the object for implementing accustomed assets in advance of agriculture removing the enclosed disparities in the state.

Role of conventional knowledge

The most important function of traditional knowledge is that general norms can be evolved for similar ecological systems across the world. For centuries, farmers have planned agricultural production and conserved natural resources with the instruments of indigenous knowledge (IK). The development of IK systems, including management of natural environment, has been a matter of survival to the people who generated these systems. Such systems are cumulative, representing generations of experience, careful observations and trial and error experiments. Indigenous knowledge is stored in people's memories and activities. It is expressed in stories, songs, folklore, proverbs, dances, myths, cultural values, beliefs, rituals, community laws, local languages and taxonomy, agricultural practices, equipment, materials, plant species and animal breeds. Indigenous forms of communication are important to local level decision making processes and for the preservation and spread of conventional knowledge. This body of knowledge has developed over generations through the process of man environmental interaction and its continuity depends on its transmission and the ability of the young generation to acquire and practice it. Indigenous knowledge systems in traditional Africa have been used by communities to protect natural resources from unsustainable exploitation thereby averting disasters that may have occurred from such exploitation. Example, Agriculture in Uganda where there is agricultural sector accounts for about 90 percent of exports and 80 percent of the employed household population. The sector provides most of the raw materials to the mainly agro-based industrial sector comprising coffee hulling, cotton ginning, tea processing, sugar production, textile mills, soap industries, edible oil industries, cigarette manufacturing, grain milling, meat processing, beverages and breweries industries, dairy and leather products manufacturing (MFPED, 2002). There are about 3 million smallholder farm households of which 80 percent have less than 4 ha of farmland and the hand-hoeing is the predominant technology for cultivation. The main traditional cash crops grown are coffee, cotton, tobacco, tea and sugar cane, while the traditional food crops are mainly maize, beans, cassava, potatoes, sweet potatoes, groundnuts, bananas and finger millet. Food crop production dominates the agricultural sector, contributing 63.8 percent of agricultural GDP (MFPED, 2003).The agricultural sector in this country is characterized as semi subsistence with low input and low productivity and the challenge facing. Uganda is to provide the necessary support services to turn widely dispersed small-scale subsistence agriculture into an engine of economic development. The sector has high potential if improved agricultural technologies are adopted by farmers besides their traditional methods of agriculture. The Government of Uganda has embarked on a Plan for Modernization of Agriculture (PMA), in order to move the predominantly rural economy out of poverty and to improve the welfare of the majority of the population. Under this policy, the focus is on increased and sustainable food production by improving agricultural techniques and practices, optimal use of land and water resources, increasing production of high-value commodities and products and gradually moving towards commercial agriculture in collaboration with the private sector. The key issue targeted by PMA and implemented by national agricultural research system (NARS) is increasing agricultural activities. The departure point is envisaged as generating appropriate agricultural technology options for transfer to the small scale and resource poor farmers. Recent developments National Agricultural Research Organization (NARO) of Uganda is mandated and conducts agricultural research on behalf public. It has developed and disseminated improved agricultural technologies to increase agricultural productivity while simultaneously preserving the natural resource base. Surprisingly however, farmers have taken up few of these while the rest remaining are un-used. This can be attributed to limited integration of agricultural practices and natural resource management practices of farmers into formal research. Many questions have been raised as to why the adoption rate of these technologies has been low. But whatever the reason, the major issue is that farmers have been previously left out in technology generation and development process. There have also been some discussions on indigenous technologies as a way of getting to understand farmers' needs. Indigenous knowledge (IK) is ideas, beliefs, values, norms, and rituals, which are native and embedded in the minds of people. It is local knowledge which is unique to a given culture or society. Indigenous knowledge is the systematic body of knowledge acquired by local people through the accumulation of experiences, informal experiments and intimate understanding of the environment in a given culture. Local people, including farmers, local artisans, and cattle keepers are the custodians of IK systems. They are knowledgeable about their own situations, their resources, what works and what doesn't work, and how one change impacts other parts of their system. IK is dynamic and it changes through creativity and innovativeness as well as through contact with other local and international knowledge systems.

Indigenous knowledge

Overall objective of indigenous knowledge system is to improve agricultural productivity of farming communities by integrating indigenous knowledge into research thereby enhancing adoption of agricultural technologies. These knowledge systems represent mechanisms to ensure minimal livelihoods for local people. Indigenous knowledge systems often are elaborate and adapted to local culture and environmental conditions tuned to the needs of local people and quality and quantity of available resources. The terms traditional knowledge, indigenous technical knowledge, rural knowledge have been used interchangeably. The need for stakeholder ownership is now well established and involves ensuring widest possible participation of those who are supposed to be the beneficiaries of the projects so that they drive the process, which includes planning, design, implementation and monitoring of agricultural system in a society. A farming systems and livelihood analysis survey was undertaken and many issues emerged from the study concerning farmers' coping mechanisms based on indigenous knowledge use at the different stages of agricultural production processes in Uganda. Most of these could be incorporated into research to enrich the agricultural technology development process and make it more relevant for the farmers. The promising indigenous farmer practices and knowledge need to be tested validated and where necessary improved. Bulindi Zonal Agricultural Research Institute (ZARDI) in Uganda, therefore carried out a study in Hoima, Masindi and Kibale districts of Uganda to investigate skills, practices and methods of crop and livestock production using local knowledge. It was aimed at identifying appropriate and promising farmer practices and recommending them for integration into conventional research. Integrating indigenous knowledge into research was expected to enhance adoption of agricultural technologies generated by the research organization that has been relatively low. The study targeted the rural people as primary stakeholders to contribute to the programme of integrating farmers' knowledge into conventional research by sharing their knowledge and insights in indigenous practices. Rice is a new crop which farmers thought had high potential of becoming a major cash and food crop in some districts like Jinja. The main vegetables grown in order of importance were obugoora, tomatoes, dodo (*Amaranthus* sp.), egg plants and cabbage. Onions, African egg plant (enjagi), green pepper, sukuma wiiki, eyobyo (spider weed like *Gynandropsis gynandra*), and pumpkins were also grown. These vegetables are also for both cash and food. The vegetables were all local varieties except tomatoes and cabbage. The main fruits grown are guavas, oranges and tangerines. Jack fruits, pawpaw and mangoes were mainly grown for domestic consumption while passion fruits, pineapples and avocados were for sale.

Crop production constraints

There are four major constraints in crop production in order of importance were recorded as lack of knowledge on improved farming methods, inadequate extension services that is one agricultural officer could hardly suffice for the whole sub-county, lack of improved crop varieties and markets coupled with low producer prices and unscrupulous middle men who in most cases exploit farmers in many ways. Other constraints of importance were crop pests and diseases, lack of capital to purchase proper farm implements thus many farmers are condemned to using hand hoes. Another challenge was unpredictable or unreliable weather conditions characterized by too much or less bimodal rainfall very uneven and erratic being intermittent drought. Farmers lacked capital to buy inputs like spray pumps and chemicals. Many farmers are unable to access loans due to high interest rate charged by banks.

Limited land to expand farming, declining soil fertility and high presence of vermins (baboons, monkeys and rats/rodents) were also some of the challenges to crop production. The livestock considered most important irrespective of communities. Almost all households kept chicken and goats for both home consumption and income. Piggery is a relatively new enterprise that has become an important source of income to a number of households. Other types of livestock reared on small scale both for sale and home consumption were cattle, sheep, rabbits, ducks, and peacocks. The main constraints in livestock keeping in order of importance were diseases, the expensive feeds, thieves, wild animals, lack of improved breeds and lack of training in modern livestock keeping techniques.

Use of indigenous knowledge in farming

Farmers employed various indigenous practices most of which were cross cutting amongst the crops grown. Early planting is one of the pillars for both indigenous and improved farming methods practiced. This is especially important in this agro-ecological zone where agriculture is rain-fed. Farmers take advantage of the early rains which also reduce the incidences of pest and disease leading to high yields. Farmers used to burn grass or trash in their farm, they prefer to plant or sow green vegetables and millet in it. Ash is assumed to be a source of nutrients and also burning is believed to kill crop pests. Farmers plant lab-lab around their farms as they believe it prevents night dancers from intruding in their gardens and other related cases of witchcraft. Farmers also practice selection of clean planting materials as in the case of formal research to control pests and diseases. In case of cassava, they ensure that the cuttings are not damaged prior to planting and that nodes face downwards to encourage effective sprouting and root growth. Farmers relied on crop rotation to rejuvenate the soil. Many farmers in tobacco growing area prefer to plant root crops in plots that originally had tobacco to specifically control pests. Farmers use rudimentary post harvest handling techniques. For instance many root crops are highly perishable crop when harvested. Farmers harvest cassava and not all of it is consumed or sold, the fresh tubers are buried in moist soil measuring from one feet deep. According to farmers, the tubers stay fresh for up to seven days. The common methods of processing cassava involve peeling, slicing, drying and storing in baskets. Bitter cassava varieties, it is peeled, sliced and left to ferment indoors for three days. It is later dried, chopped and finally ground or pounded into flour. Fermentation was reported to reduce the cyanide level in or the bitterness of the cassava, a process also recommended by conventional research. Grain crops like beans, farmers ensure that beans are planted as the second crop in the rotation system. Broadcasting the seed before ploughing is still a popular method used when planting. Early plating is also preferred to allow crops receive enough rainfall and reduce pests and diseases incidences. Farmers also used concoctions of ash, goat droppings, dry and water as insecticide to control pests and diseases. Ash is also used as an important component for preserving specifically beans. Some farmers mix ash with water to form a light paste and add to the beans before storing in containers made from dry banana leaves, fibers and sticks. Currently however, this method is no longer common because such storage facilities needed to be kept outside and there have been rampant cases of theft. Use anthill soil as a preservative during long storage. Farmers also put elephant grass flowers locally known as etete and leaves of obukomera and neem tree in beans because the scent produced by the plants species repels storage pests. Caterpillars were reported as the most important sweet potato pests. They are controlled by picking the infested leaves and burying them. Ash is sometimes mixed with human urine and sprinkled on vines. Farmers sometimes apply cow dung on potato vines and also rogue potato leaves. Lastly, farmers also carry out some form of rudimentary biological control of pests. They catch black ants and release them in sweet potatoes fields to eat the caterpillars. Farmers know which crops require fine seed beds and which ones do not and when they come in the rotation system from years of experience. Banana suckers with weevils are rouged and pseudo stems split and placed upside down on the stools to trap weevils. A mixture of human urine, leaves of omuhoko, some water and pepper is fermented for 14 days and applied around the banana stool to control weevils. This mixture is placed at least 2 feet away from the banana stool. Urine fermented for 14 days and mixed with ash and omuhoko controls banana weevils. Urine is applied directly on the banana stool to kill banana weevil. This partly explains resistance of banana stools to pests and diseases near kitchen since the kitchen is always a source of ash and its compound particles. Farmers mix banana varieties when planting to control pests/diseases. This practice is different from current practice where each variety is planted in sole plot for commercial purposes. Banana for juice are placed in Embiso and Itunguru (a structure constructed above the fireplace) to ripen. The Bokora/Kivuvu type of bananas is cooked and sun dried when they ripen. It can store for several weeks and mainly served with tea at break first. Human urine is fermented for two weeks and it is mixed with water at a ratio of 1:2. If the mixture is being prepared to "spray" tomatoes, cabbage leaves are mashed and added to it. The mixture is being prepared for "spraying" cabbages tomato leaves are mashed and added to it. A mixture of fresh urine and water at a ratio of 1:1with marigold leaves, pepper, and tobacco leaves are added to the mixture and kept fermented for 2 weeks. This concoction is mainly used to increase vegetables and bananas production booster with approximately three liters is applied per banana stool. Farmers expressed reasonable knowledge on soil fertility indicators. This is mainly determined by soil characteristics, types of weeds growing in an area and crop characteristics such as yields, vigor and the general appearance of the plants. Soil types, degrees of soil fertility, and land use categories were also desegregated by farmers. Color, texture and even taste usually distinguish soil types while some classified their soils based on vegetative cover. 8 Farmers can determine that when the soil is exhausted. The water retention level reduces substantially (becomes too porous), the plants/ crops or weeds are stunted and yellowish and the crop yields decline at an increasing rate. Decline in soil fertility is also determined by evident loss of crumb structure. It also becomes very light, dusty or sandy in texture. The soil color changes from dark or brown to reddish with increased number of stones. Soils are exhausted; it allows the growth of weeds like wandering jew, spear grass, *amatojo and amaranga.* The weeds like bidens pilosa, embuura, milk weed (euphorbia species) also indicate soil infertility. Farmers use various means at every stage of the plant growth to cope with the problems of soil fertility loss. This include making mounds by collecting and heaping trash in preparation for planting sweet potatoes, potatoes, tobacco and vegetables which are preferably planted on raised seed beds. Other common practices are fallowing and planting elephant grass at the beginning of fallow period, for one season for up to 4 years depending on land availability. Intercropping beans and maize, groundnuts and maize, millet and maize and agro forestry is practiced especially for fruit trees and coffee to improve soil fertility. Crop rotation where cassava is planted as last crop in rotation is also a popular practice as many farmers believe that cassava is not a heavy feeder and that when cassava leaves wither and drop, they decompose and add manure to the soil. Farmers heap soil and trash around the plant while weeding and make bands. Water runoff and/or soil erosion is reduced using soil bands *(fanya chini and fanya juu).* Farmers also add crop residues like kitchen waste/refuse and manure from goats, chicken, and cattle to their fields and fallow plots to enhance nutrient status. Farmers mostly use elephant grass and maize stalks to conserve soil moisture and add manure after decomposition when mulching. Mulching was practiced in at least 4 out of 10 households in every village. Many farmers had stopped the bush burning practice and were making and applying compost manure. Indigenous knowledge was mainly applied to crop production than in livestock rearing. Farmers mainly used for livestock, on chicken and goats than to the rest of the animals and birds. There was little information on piggery as far indigenous knowledge is concerned probably due to the fact that piggery is a relatively new enterprise in most households.

ITK in chicken production

There were interesting findings on use of indigenous knowledge in chicken production and some reported below, a brooding hen abandons the eggs completely because of pest infestation like mites, farmers put dry banana leaves in a mortar and position it near a fireplace to aid the eggs brooding process. The eggs are regularly turned and eventually the chicks hatch; Farmers prepare and place a small bottomless basket to encourage hens to lay more eggs and hatch many chicks. The baskets are placed in a small round hole lined with dry banana leaves farmers supplement feeds for hens with millet and maize to encourage the hens to lay more eggs. The eggs are removed regularly in order to increases the number of eggs laid. Chicken are fed on a mixture of millet and paraffin to prevent sotoka (coccidiosis). This however is believed to make the hens take longer to lay eggs. Chicken infected with coccidiosis are injected with or given mixture of ash, ground pepper and water to drink. Cannabis is mashed and added to water and used for treating coccidiosis. Farmers also make an incision under the chicken wing to prompt bleeding from the vein of an infected (with coccidiosis) chicken. The vein is usually distinct i.e. dark/black in colour. Farmers give chicken insects from empike (enkubebe) to peck in order to control New Castle Disease. The canabis leaves are also pounded and added to water, boiled and given to the chicken and other birds. The cannabis can also be given to chicken to peck for the same purpose. Tiny fish is fed to the chicken to control and to treat the New Castle Disease. Green leafy vegetable is mashed, added to water and the mixture given to chicken to drink and to destroy pests, farmers apply paraffin and vaseline on affected area especially around the chicken eyes. An old tyre is burnt in chicken coop to destroy mites. Farmers also cut and place a moist tree branch in the chicken coop to attract and trap which is removed and thrown far from the homestead when it is laden with the pests. In tobacco growing areas, farmers spread fire cured tobacco leaves on the chicken coop floor to kill mites. Farmers give petero leaves to chicken to peck as a de-worming agent. It was also reported that the same leaves are used for de-worming humans. Pawpaw seeds also are given to chicken as dewormers. Farmers use their hands to aid goats experiencing difficulties while delivering. In case the placenta to disengage, farmers give the goats cowpea leaves. Others prefer to hit the bottom of the goats with brooms ease the release of the placenta. Another remedy is giving goats a mixture of enderema, water and salt to aid the disengagement of the placenta after the goat delivers. A pregnant goat fails to deliver, it is tied near a fireplace to presumably burn fats believed to block delivery. Worms in animals are treated using green vegetable with a specific measure that farmers know. Diarrhea is treated using a mixture of cowpea leave and little salt. The mixture may either be boiled or not. Potato leaves are also used to treat diarrhea in goats. A quantity of 0.13mls is administered to each goat using a syringe. Farmers did not give an explanation of how they use soap or foam to prevent fever in goats although they said they use it. In addition, fresh cannabis leaves and little salt is mashed and added to water to treat goats suffering from fever, cough, poor appetite and/or diarrhea. To increase milk production, farmers add salt to banana peels, bean soup and leaves squeezed with salt and little water and feed it to the their animals to increase milk production.

Another common practice is the application of sap from the rukoni tree on the infected spot. Others 10 believe that tying ekinyege around the neck of the infected animal treats ECF. Mastitis on the other hand is treated by directly pressing a hot metal on the affected areas. Farmers use a tedious approach of pricking fleas (embarabara) with a needle and leaving them to die on the animal to control the pest. Some livestock experience constipation and in case of such a condition, farmers mash leaves of omukyora or ekibirizi with some salt and give it to a suffering animal. Sources of indigenous knowledge are identified from different sources. Main sources were reported as interactions with the elderly, parents, grandparents, relatives and friends. The commons sources of indigenous knowledge are through visits where one finds a technology being applied and picks interest in it. Migration of people from other parts of the country with different ethnicity, radio programs and extension workers and own discoveries were noted as supplementary sources of indigenous knowledge. Constraints of using indigenous technical knowledge is education and exposure especially of the young generation to modern training have biased people's attitudes towards using indigenous knowledge. Some farmers feel it is time consuming, exhausting and sometimes dangerous to hunt for herbs in the forests and bushes indigenous knowledge was reported as not effective in large scale production. Some religious beliefs do not encourage traditional beliefs and technologies regarding them as demonic and superstitious. Regressive government laws prohibiting some practices like growing cannabis, unauthorized tree cutting and rudimentary castration of livestock. Depletion of most trees and herbs that are sources of local medicine and lack of standardization and documentation of indigenous technologies and practices variation in prescription such as quantity to administer at a time and for how long the educated people despise some indigenous methods referring to it as ineffective and dirty. Selfishness that inhibits people from passing on knowledge to others advantages of using indigenous knowledge in agricultural production ITK products are cheap and in most cases cost free in monetary terms.

Advantages of indigenous knowledge practices

1. Indigenous knowledge creates social harmony and cohesion;
2. Indigenous knowledge is easy to grasp the concepts and practices because knowledge can be passed on orally using the local language they are available domestically like urine, ash and hot pepper;
3. Indigenous knowledge is not harmful to human health;
4. Indigenous knowledge does not have side effects;
5. ITK products are prepared and obtained on demand and problems associated with expired of products are not of much concern.
6. Application of ITK products does not always demand specialist attention like veterinary doctors or extension workers.
7. Jhum cultivation
8. Farmers of North-East are not trained for participatory research and they term traditional knowledge as unscientific. Before making any suggestions, they should consider that climatic conditions and topography differs from region to region. In the Northeast it rains very

heavily. Therefore, the region is unsuitable for terrace cultivation which has been suggested as the alternative to *jhum* cultivation. It also requires a lot of labour and, hence, has proved

Conclusion

Indigenous technological knowledge provides admired inputs to accomplish able use of accustomed assets and extends accordant abutment for adequate development. Indigenous techniques acclimated in altered basic of agriculture adjustment are mostly organic, eco-friendly, sustainable, applicable and amount effective. But, there is a charge to explore, verify, adapt and scientifically validate these practices for their added use and application. Indigenous knowledge plays a capital role in adequate grassroots innovations. Such grassroots addition in large quantities differs beyond altered sectors with account to the characteristics, sources, actors complex etc. Indigenous knowledge is not discouraging bare for altered grassroots innovations those are different, which in about-face decides the captivation of truthful set of actor. In case of sufficient societies, the surrounded original alone is the above the performer. In abounding cases, the indigenous communities are not able-bodied familiar of the amount of their indigenous knowledge which has been casual from bearing afterwards generation. The truthful institutions and non-government organization could play sensitive role in these recommendations for accommodation structural design amidst the indigenous affiliation and popularization of adequate methods and techniques. In today's context, there is a burning charge to appraise and popularize aboriginal innovation. Government schemes and analysis and development activities should knowledge indigenous innovators. The cardboard precise discarded a few examples of indigenous knowledge are acceptable anywhere. There are wealthy added such examples an allotment of the indigenous groups in all the countries. As a lot of the sufficient knowledge and technologies are undocumented, there is as well a charge for added analysis in this field. Otherwise, this admired knowledge will be abolished in the abreast future. There should be an able accord amid the indigenous knowledge and modern knowledge. Such display of accord is occurring in some of the sectors such as ethno-medicine. The aforementioned should appear to all added sectors so that the contributions of the indigenous innovators appropriately knowledge the society. Especially, there should be able accord amid indigenous knowledge and modern technological knowledge. However, such activity is complicated to a above admeasurements due to which there is a synopsis of able affiliation amid the conveyance and accomplishing of the indigenous and modern knowledge. An adapted affiliation amid the adequate and modern knowledge and technology systems has immense popular to account the present-day society.

Some pertinent short questions

1. What are the possible documentations of ITKs?
2. Write down five possible of stakeholders of ITKS;
3. Give brief note on role of ITKs in crop production;
4. State the role of ITKs in restoration of ecology;
5. What are the roles of ITKs in maintaining environmental equilibrium?

10

Plant Disorders, It Causes, Symptoms Treatment Materials and Plant Protection in Ancient India Through Indigenous Traditional Knowledge (ITKs)

Introduction

Indigenous Technological Knowledge (ITKs) refers to the knowledge that acceptable to all as absolute hearing and developed about the meticulous altitude by the people irrespective of gender that is women and men ancient to an a specific geographical area. This ancient technological knowledge that humans in an accustomed association accept developed over time and abide to advance it, is predicated on animal kingdom adventures on accumulation scale, activating and changing, activated throughout a lot of cases over centuries of use. Highest accessible knowledge to bounded knowledge and ambiance and put greater weightage on abbreviate risks instead of maximizing profit. The traditional technological knowledge (ITKs) covers a complicated sphere of subjects, *viz.* crop production, livestock rearing, accustomed knowledge management, ailment preparation, healthcare, insect administration, and others. The utilization of non-chemical methods for protecting lives and crop aegis is already accepting in several countries including India. The administration strategies developed and answer by the Governments is now supported the utilization of plant extracts. If an accomplishment is fabricated arise the cultivation of Indigenous Technological Knowledge (ITK) based articles on the small scale, it is often an economically applicable advantage for the appropriate development of eco-friendly chemicals.

Indian agriculture has been started over 10,000 years old, over millennia, farmers developed innumerable practices to auspiciously abound crops and accession animals within the awful adapted agro ecological regions of the Indian subcontinent. About 170 years of acceptance brought in new techniques, some advantageous in addition to a few harmful event of ancient Technological knowledge (ITK) systems, including administration of accustomed environment, has been a bulk of adaptation to the humans who generated these systems continued aback. Million of ITKs were supported to avail knowledge of actual and living assets to construct sure basal livelihoods for bounded people.

Indigenous Technological Knowledge (ITKs) has immense popularity for innovation, acceptable eco-friendly agriculture at the grassroots level. India might be a country busy by a bulk of ancient communities, hundreds of which accept their own set of the altered acceptable knowledge in their respective domain. Amongst of these data and technologies available are if not at par with the knowledge and technology adjustment these were accommodated to able after-effects in symbol to forward-thinking technologies arise accumulation agricultural great quantity and self-sufficiency. These acceptable knowledge and technologies played a convincing role within the all-embracing socio-economic development of the communities. Findings on a number of the traditional knowledge and technologies, with adapted advertence to the findings of Indigenous Technological Knowledge (ITK), accustomed amidst a bulk of ancient communities were agitated out and therefore the acceptance of the aforementioned in addition has been evaluated. A complicated domain of various sectors including agriculture, beastly husbandry, fishing, and bolt was advised for the aim as of these sectors are acute surrounded by the Indian context. It has been empiric that there's a burning charge for its affidavit and to bottle the Indigenous Technological Knowledge (ITK) of specific communities, of which are at the border of extinction. There's an synopsis of able agreement amid of the conveyance of ancient and advanced knowledge. As agriculture may be a basic breadth of the digest and a suitable accord has been able during this wideness during the plan period. Total crop cultivation has added assorted and yields of above crops like paddy, pulses, oilseeds, and vegetables accept added than trebled within the tolerate limits of four and half decades. The absorbing continued appellation advance in agriculture has helped in demography the accompaniment out of famines and severe shortage into one in the middle of the surplus states within the country and ensuring food, diet and security. But although the charge of authoritative accessible an able aliment accumulation of reasonable superior for the citizenry has been proficient, the agnate charge of acceptable and open agricultural advance within the state.

Spatio-temporal variations

Indian farming, which is traveling through an alteration phase, is boring but completely adopting the agency and agency of aggravation administration for acceptable agriculture. Adoption of ITKs based crop sponsorship measures as another to pesticides knowledge advice in narrowing the biodiversity of familiar enemies, but as integrated pest management (IPM) is a knowledge-based and agriculturalist friendly approach, apprenticeship of farmers on alternatives to pesticides accept to be adapted a priority. Although successful of these acceptable practices are precise by accurate findings, yet these has not been accepted and adapted into technological recommendations to accredit farmers to decrease their costs and accomplish agriculture added sustainable. Farmers consistently apprentice and unlearn, allotment adapted knowledge in their attempt for an acceptable livelihood.

In fact, agriculture is characterized by advanced assortment and ample spatio-temporal variations in advance and productivity. Such disparities area could cause of affair for accessible reasons. In such appearance of the facts, findings on spatio-temporal variations in agricultural development is explanation pursuing. Variations in the Indian agro-climate from close to moderate altitude accept led to the development of very old abilities based on innovations at farmers' fields. Unfortunately, these abilities do not acquisition abundant appreciation, as these are not activated in the land by researchers. Although flourishing of these acceptable practices are accurate by accurate findings, yet these accept not been adapted into technological recommendations to accredit farmers to decrease their costs of inputs and accomplish agriculture added sustainable.

Present arts of knowledge

Most of the studies focused on the disparities in agricultural achievement at the supplement within the country. However, their accomplishments do not acknowledge a bright trend over time and abridgement an audible arrangement beyond states. The studies conducted several workers arise the actuality of advanced variations in large quantity and all-embracing agricultural achievement amidst the states. Someone advised the delimited development in agriculture in 15 states of India during the year 1971-2007. Their activities adumbrated that restricted disparities in agricultural performance, which added during 1971-88, chasten during 1988-2007 suggesting an affectionate of convergence. Introduction top acquiescent irrigation-responsive and fertilizers acknowledging HYVs varieties of rice and accelerated agriculture change three to four crops in a year in the aforementioned section of acreage appropriate huge volume of water. The water was supplied either credible or from arena entitle resources. As results, the burning of identify table occurred and found to be sharp. Scientists conducted three criterion surveys in contempt years (1980-81, 1990-91 and 1998-99) in specific areas. A notable award of the studies is that bordered disparities to accept and to chasten in the post-reform period 1991-1999 because of accomplishing of at the back extent development programmes by the government. Scientist advised accoutrement the year 1980-81 to 1992-93 that as well that out the achievement beyond districts with the districts and as able-bodied specific commune of western arena of Odisha that credible bigger achievement than others. The principle of adopting ITKs was advised the bounded disparities in agricultural development in altered districts and zones of different states during the criterion years 1980 and 2000. The activities adumbrated top inter-district variations in agricultural abundance with college abundance in the credible acreage areas than others. In a contempt findings regarding the agriculture is accoutrement the pre-liberalization (1971-90) and post-liberalization (1991-2008) periods, where it was found that the credible bounded disparities in agricultural advance showed admirable achievement.

The analysis of series of literatures in the accompaniment features out to a bulk of deficiencies of absolute analysis works on the subject was done. First, the accomplishments arise to be alloyed and diverse. Second, with agricultural achievement of an span afflict decidedly by its accustomed by the institutional and technological characteristics (which alter broadly amidst and aural the states), absolute analysis absorption abundantly at the accompaniment like and the change of basal and micro-level planning emphasizing the accent of commune like access to agricultural development wideness in charge for studies at the lower like becomes apparent. Third, there is an absence of studies on agricultural achievement disaggregate in the country. Besides a lot of changes in agriculture pattern, irrigation coverage, agriculture intensity, fertilizer consumption, boilerplate admeasurements of operational backing and allotment of agricultural workers accept taken abode in accompaniment agriculture during the contempt years. These changes knowledge accept afflicted the agricultural achievement of states abnormally and causeless to say they could not accept been grasped in the absolute studies. There is appropriately a charge for advantageous these analysis gaps and adorning the absolute literature.

Average abundance of agri-inputs is random admeasurements of agricultural achievement of an area/district. Since land and labour are the two important agency inputs in agriculture, advice regarding to achievement per multitude of both land and labour can be acclimated to appraise the like agricultural development. But indicator of labour abundance is a difficult assignment and whatever appraisal can be fabricated in this attention that is apprenticed to be exposed and impulsive for accessible reasons. Furthermore, the accompaniment is advised to be the poor agrarian states and characterized bottleneck in agriculture, land is the scarcest agency and it is almost simple to compute achievement per assemblage. The land abundance as a key indicator of agricultural development and such abundance are abstinent in agreement of yields of crop grains and paddy yield per hectare of restricted area. The crop grains and paddy yield accepted that have been called because they aggregate 74 % and 45.5%, respectively of gross cropped area (GCA) of the country.

ITKs in plant disease control

1. Traditional modification of crop protection that farmers accept learnt from their continued affiliation with the land, its flora and fauna were based on eco-friendly systems of adequate cultural practices and judicious circling of crops and knowledge of pests and their life-cycle;
2. In case of bunchy top diseases in chilies, blanket application of ash, spray of aqueous decompose of scorched accoutrement in joined areas;
3. Indigenous technical/traditional knowledge (ITK) is relationship is enclosed within rural in origin. It is acclimated in get used to expectancy for adequate germination, soil entitlement and soil fertility management, administration of insect aggravation and throbbing of sphere and animals, column autumn administration. There are several agriculture practices of India which are still organic in nature that altered states of sustainable, eco-friendly, applicable and cost-effective;
4. Some farmers acclimated to administer ample cow dung abreast the collar arena of chilli ball to ascendancy fungal disease, viz. damping off and dieback could be controlled;
5. In case of soil-borne disease, miserable rot and neckline rot were acclimate meeting cake, of *karanj* and *neem* as pre-eminence measures;
6. 20 kg of *Casuarina equisetifolia* leaves are uphill in water for 20 min, afterwards cooling, the band-aid should be filtered. Again the solution is infected with water and can be accustomed to superiority some bacterial and fungal disease;
7. Prepare band-aid from 2 kg leaves of papaya in 3-4 liters of water and accumulate in overnight, afterwards filtration, this is mixed with 50-60 liters of water and 250 ml soap band-aid again added to it, is able to superiority yellowish-brown disease of rice;
8. Cultivation of marigold followed by *Solanaceous* vegetable crops is able to superiority over bacterial diseases, and
9. Leaves of khair (*Acacia catechu*) can be put into water cannel to superiority yellowish-brown fragment disease of rice.

ITKs in Insect-pest superiority

The application of chemicals for exasperation of superiority leads to aliment poisoning the soil, water and ecology pollution. These chemicals are actualizing ecological laterite and comply insect-pest to advance resistance. At these moments, an indigenous knowledge accompanying to sphere protection in agriculture would play basic role.

Management of insect pests through ash: Sprinkling of ash over and about the vegetable crops like onion, okra, brinjal, amazon and cucumber and in fields is able adjoin insect pests viz, beetles (pumpkin beetle, had beetle etc.), blade defoliating insects, blade miners, thrips and aphids. It is the cheapest conveyance for small farmers. Ash used to act as a detergent. Chewing and sucking splash of insects, attainment might be difficult to chew. Identification of locations due to degradation of ash administration of insect pests of quick growing crops through like *Aloe barbadensis* (gwarpatha) + *Nicotiana tabacum* (tobacco) + *Azadirachta indica* (neem) + *Sapindus trifoliatus* Linn (aritha). The mixture is finer acclimated adjoin the insect pests of quick growing crops. Management of insect pests through cow urine + *Vitexnegundo* (Nirgundi) + *Ferula asafetida* (Hing) could also be successful proposition.

These mixtures are actual effective, eco-friendly insecticidal analysis adjoin insect pests of aureate crop. It repels the insect pests.

Management of pod borers in gram crop through whey (lassi) + *Aloe barbadensis* (gwarpatha) + *Nicotiana tabacum* (tobacco) could also be practiced;

Popular ITKs against crop pests

The Vrukshayur vedic texts as well board advice on various adequate aggravation management practices piece of equipment locally attainable resources. A deluge of abstract is attainable on the role of cow urine adjoin assorted insect pests. It has been approved that botanicals brewed in cow articles like cow dung and cow urine could beautify microbial cultures which helped in accouterment sphere nutrients in accession to acting as annoyance repellents. Application of organic wastes based on the methods recommended in Vrikshayur veda offers amazing befalling to advance novel, eco-friendly methods of manuring and ball protection through research.

Amongst the practices of accessories and bait can be indigenously able application of locally attainable knowledge for bigger ecology and superiority over the insect pests. A few examples are cited below;

1. Trapping rhinoceros beetles (RB) in coconut: a mud pot with three house of it is to be abounding with water and to this 250 g of delicate skirmish chunk is added. The pot is again alive in the clay with its aperture in

line with the soil. That appears to smell of the block attracts the beetles which abatement into the water. Just 2-3 such pots in one hectare of acreage can ablaze beetles from the area. Slices of ache angel are as well acclimated to allure RB. In an annular artificial 2 slices of pineapple are taken and an avenue aperture is fabricated to acquiesce the rain water to drain. The beetles are admiring towards the pineapple and get trapped.

2. Trapping red approach weevil (RPW) in coconut: The mid rib of attic blade is cut into baby pieces and crushed, abode it in an earthen pot either with 1 lit of water 100 g jaggery and 10 g tobacco crumb or with sugarcane molasses 2½ kg or booze 2½ litres acerb acerbic 5 ml aggrandize 5 g. Another pot with aperture at its basal is placed over it. This adjustment is fabricated at 3-4 corners of the attic orchard to allure and allurement the beetles. The admixture of jaggery, tobacco and water is to be added already in an ages in case the in the air allurement is taken up;

3. Trapping bake-apple flies: Bake-apple fly (Dacus dorsalis and D. cucurbitae)incidence) is commonly apparent in mango and cucurbits. A bargain bake-apple flies allurement to activity this insect annoyance can be fabricated as follows;

   (a) 20 g of Ocimum atelier (holy basil) leaves are ashamed and the extract forth with the mortified leaves are placed axial a top story shell, which is again abounding with 100 ml water. To access the befitting superior of the extract, 0.5 g citric acerbic is added and the extract is again berserk by bond 0.5 g carbofuran 3G. The accessories are relevant from mango timberline branches at an amount of 4 accessories per tree. The bake-apple flies augment on the osmium abstract and are killed.

   (b) Accomplish a allurement application of a 2-liter disposable water bottle: Two holes at a acme of 5 cm from the basal of canteen are fabricated and for blind the trap, use a cord which is pushed through a aperture accomplished in the centre of the cap from inside. The attractant admixture for bake-apple flies is again able by bond 1 cup of vinegar, 2 cups of water and 1 table spoon of honey and all-overs this able-bodied afore use. Flies will access the alembic and abatement into the attractant.

   (c) Trapping sucking insect pests: Ablaze chicken adhering accessories are acclimated for monitoring/controlling aphids, thrips and whiteflies. While, ablaze dejected accessories can abandoned be acclimated for ecology thrips and ablaze white adhering accessories for flea beetles. Set up adhering accessories for ecology whitefly, thrips etc. @ 10 accessories per ha. Locally attainable abandoned tins can be corrective chicken / coated with white grease / vaseline / brush oil. Position the accessories at 50-75 cm uphill the plants. Alternatively, chicken water pan accessories as well accepted advantageous for simple population counts of aphids based aloft which insecticidal ascendancy can be initiated.

(d) Trapping swelling beetles: Dejected containers, abounding with water with little bactericide are claimed to allure abscess beetles;

(e) Trapping moths: Mix 500 ml of aloe extract, 1 kg of brush chunk and add acrylic as adhesive. Put this in a mixture in an advanced opened disposable container. Apply in cardinal locations of the land @12 kg/ ha;

Control of slugs in kitchen garden: Set the rinds of grapes with a little bright larboard axial with upside down (like an igloo-style) in kitchen garden. The slugs will adumbrate under the grapefruit and die.

Mechanical control practices

Attracting birds: Erection of bird perches @ 25/ha facilitates predation of abecedarian stages of insects.

Bait for ant: All-overs generally assure honeydew address *bacilli* such as aphids, mealy bugs, and scales from advance by accustomed enemies. Sometimes all-overs move these honeydew-producing insects from ball to plant. Ascendancy of all-overs generally leads to added able biological ascendancy of sucking pests. The allurement can be fabricated by abandon 1 tea spoon delicate boric acerbic acid and 10 tea spoons amoroso into 2 cups of water; this admixture can again be captivated into affection assurance which are larboard abreast ant trails.

Gundhi bugs in rice: Fix asleep crabs, frogs or even pieces of jackfruit (*Artocarpusheterophyllas*) to bamboo sticks and abode them in rice fields afore alabaster stage. This will allure gundhi bugs and accumulate them alive till the chef date is over.

Rat / rodent control: Fruits of *Mucunapruriencs back* (Fam: Papilionaceae) are kept in the alive rat burrows. If the rodent enters into the hole, it collides adjoin the bearded fruits with acid hairs and blade the atom with irritation. A mixture absolute 90% sesame or groundnut or niger scrape with 5% blubbery amoroso crystals and 5% delicate ball or tube is placed in a basin alongside each other rat holes and if rats augment this mixtures they die aural a week.

Inserting 10 – 12 inches continued alpha pieces of axis of *Jatropha* ball into alive rat holes makes the acreage rat charge less (Kanojia, et al.,2005)

Use of botanicals: Botanicals are readily attainable than bartering articles as they abound in the bounded environment. Reviving and modernizing age-old agriculturalist conveyance through the enhancement of ethno botanicals has apparent that farmers are added adequate appliance ball abstracts than bartering bolt and those botanicals can activity a agnate like of ascendancy if assertive guidelines are followed to their use.

Aloe (*Aloe barbadensis* Fam: Aloeaceae) vitex (*Vitex negundo* Fam: Verbenaceae) extract: Soak vitex leaves (5 kg) in 10 liters water. Remove the alien allotment of the aloe leaves (2 kg) and bull work in water to get the extract. Mix the two extracts and adulterate in 50-60 liters of waters to awning 0.4 ha area. Add 50-60 ml soap in the admixture and aerosol aboriginal in the morning or backward in the afternoon. This *Aloe vitex* extract is appear to superiority armyworm, bearded caterpillar, rice blade folder, rice axis borer, semi-looper, bacterial and fungal diseases.

Coriander (*Coriandrum sativum*) for spider mite control: Coiander acts as a hateful and to become accustomed the extract swelling 200 grams of drove seeds in 1 liter of water for 10 minutes. Adulterate extract with 2 liters of water application in the morning on infested ball locations to superiority spider mites.

Marigold and chilli extract: Chop 500 g of accomplished marigold ball and 10 hot chilli pods; Soak them brief in 15 liters of water, make impure the clarify with water at 1:2 ratios and add soap @ 1 tsp per liter of extract. This controls a lot of agricultural pests.

Turmeric (*Curcuma domestica*): Soak disconnected miserable (20 g) in 200 ml cow urine. make impure the mixture with 2-3 liters of water and add soap (8-12 ml) and spray. The extract controls aphids, caterpillars, red spider mites and arenaceous bane.

Indian privet timberline (*Vitex negundo*; Fam.Verbenaceae): Soak 2kg vitex leaves brief in 5 liters of water and abscess the admixture for 30 minutes. Add 10 liters of water and soap (10 ml) and spray. This controls bearded caterpillars, rice blade folder, rice axis auger and semi looper.

Neem leaves extract: Pound 1 kg neem leaves and habitat it in a pot with 2 liters of water. Cover the aperture with bolt cocked and leaves it as such for 3 days. Make impure the extract at 1:9 with water and add 100 ml of soap before spraying. This controls aphids, grasshoppers, blade hoppers, ball hoppers scales thrips weevils and beetles.

Calotropis gigantean: Leaves are preserved in big earthen pot abounding with water for two weeks. The water is activated @ 0.5 l / timberline to ascendancy termites. The leaves are replaced afterwards 2 months.

Formulation of control measures

Cow milk: Cow's milk appeared to acts as an accomplished sticker and spreader due to attendance of casein protein has accomplished spreader and sticker property. It can be acclimated @10% aqueous abeyance for finer authoritative farinaceous mildew. Milk sprays induced systematically acquired attrition in chilli adjoin blade curl, a viral diseases.

Cow urine and dung

(i) Cow urine adulterated with water in adjustment of 1: 20 is not abandoned able in the administration of bacilli and insects, but as well acts as an advance apostle of crops

(ii) Cow urine accept been begin able adjoin abject bugs, thrips and mites (Peries, 1989) and adjoin column alpha insect pests of cowpea. Drove 5 kg neem leaves in water, add 5 lit cow urine and 2 kg cow dung agitate for 24 hrs with alternate stirring, clarify the abstract and adulterate it in 100 lit of water for spraying over one acre. This abstract is advantageous adjoin sucking pests and abject bugs.

(iii) In brinjal, application of cow urine 10% starch 1% either discarded or on the other hand with chlorantraniliprole 18.5% SC insecticides was found to be effective amount.

(iv) Botanicals brewed in cow urine/cow dung: The cow urine decoctions of botanicals accept been appear as able adjoin the assorted insect pest's afterwards apparent adverse aftereffect on their accustomed enemies.

(v) Cow urine 5% with neem berry particle extraction at 5% and cow dung 5% showed anti-fee dent and anti-ovipositional furnishings adjoin Helicoverpaarmigera; (vi) Amongst 14 cow urine alloyed botanical extracts activated *Lantana camara* Linn and *Vitex trifolia* were appear able adjoin aphid, **Lipaphiserysimi**. Crude extract of Daturaalba (20%) cow urine (20%) was able adjoin axis auger and blade binder in Basmati rice.

(vii) Barapatre and Lingappa (2003) as well accurate the cap knowledge of cow urine forth with assorted botanicals *viz*., NSKE, Pongamia, Vitex and *Aloe vera* adjoin S. litura and H. armigera in groundnut and chickpea, respectively;

(viii) Combination of cow urine with NSKE and Vitex bargain the shoot fly bane in sorghum.

(ix) Cow urine brewed karanj leaves (10%) / neem leaves (10%) were ideal in account of bankable bake-apple crop (135.5-141.7 q / ha) and benefit: amount adjustment (38.20:1 – 42.68:1) admitting of their knowledge adjoin the shoot and bake-apple auger in brinjal.

(x) Ash: A blubbery band of ash is either advance on the clay about plants or brindled on foliage to assure it adjoins an array of pests. Besides acting as a concrete adulteration ash on crop foliage interferes in the actinic signals appearing from the host plants appropriately obstructing the antecedent host area by pests. Ashes from burnt approach blade and bunches accept been commonly acclimated in the eastern locations of Nigeria to dust the leaves of okra to assure adjoin blade bistro beetles, *Podagrica spp*. Application of ash @ 50kg/ha kerosene 5% and spinosad 45% SC generated best account amount adjustment of 4.8:1in brinjal.

(xi) Kerosene: It is readily attainable with the farmers and can be acclimated with soap instantly to abolish the insect pests at the alpha of beginning bearings and afterwards the desired/recommended strategies may be followed. The use of Kerosene-soap-water chrism has beforehand been appearing as an acquaintance insecticide for acute and sucking insects. Similarly, the account of this chrism adjoins calibration insects, bugs, mites, aphids and blade miners have been reported also). It also appeared that the cap knowledge of SABRUKA (an admixture of soap, water and kerosene) adjoin insect pests of cowpea in the arctic Guinea Savanna found effective. Kerosene exhibits phytotoxicity at seminary concentrations and therefore, its use as foliar spray should be belted up to 1 or 2%. Adapt a 4 lit banal band-aid of soap kerosene admixture in the under accustomed proportion;

(xii) 3.5 liter of water 48 g soap (1.2%) 500 ml kerosene (12.5%) before spraying infect 250 ml of this mixture with 4 liters of water. Oils may as well repel some pests but the botheration of phytotoxicity cannot be ignored. Visible leave damage, or added cautiously abridgement in crop could be possible. Bi-weekly oil applications bargain whitefly counts on leaves by two thirds, but crop on the oil-treated plants was as well bargain compared to basic plants. Five oil sprays controlled farinaceous bane in grapes but bargain amoroso levels .

(xiii) Milk treatment: condensation of viral diseases has been begin to activity mostly in contaminated plants of *Solanaceae, Piperaceae and Malvaceae* families by milk spray. It was activated that alpha milk aerosol reduces about 73 per cent of tungro mosaic virus infections in tobacco. Brewed buttermilk (11 in 20 liter of water) sprayed on tobacco plants cures blade coil disease. Alpha dupe milk sprayed over plants like chilli (*Capsicum annum*), brinjal and breed like atramentous pepper as well helps superiority bake-apple and leave curl. The near-neutral pH of milk not abandoned washes off the fungal spores but the adhering acreage of milk fat as well prevents formation of spores.

(xiv) Neem and ash: Before clearing paddy, seedlings are kept in baby plots of formed water alloyed with ash and pulverized neem seeds. For a artifice of 15 sq. feet, bisected a kilogram of neem seeds and a kilo of ash is adequate for bond with water to board 50 bundles of bulb at a time for a year of one hour. Advised bulb aftermath a crop charge less of annoyance and diseases.

(xv) Lantana leaves: Leaves of lantana, an abhorrent weed, can be utilized to bottle and activity candied potatoes. Insect advance can be belted to a bald minimum with no use of poisonous insecticides.

(xvi) Inducing conception: an indigenously developed finding of giving analeptic to dairy animals is accepting popularity. The cows, if fails to conceive, an analeptic fabricated of the case of Cassia fistula, breakable leaves of jackfruit tree, wretched of *Plumba gozeylanica* and leaves of *Cleroden druminerme* are powered and the admixture is fed to animals forth with jaggery.

(xvii) Deworming: reddish-brown crumb calm from the bake-apple of the broadly developed kaamal timberline (Mallotus philippinenis) is acclimated for the analysis of bastard infection and or ache in beasts and buffaloes. Depending on the blazon of ailment, 50-60 g of kamila crumb alloyed in blood-warm water/butter milk is administered orally.

(xviii) Udder swelling: A few canticle afterwards calving, dairy animal of sudden stops bribery due to swollen of the udder. To pleasure this, disintegrate quartz stones are out cold through a muslin cloth, alloyed with adulate and anointed over the udder. It is activated alert a day till the animal beastly is cured.

(xix) Treating fractures: A adhering for alleviative a burst cartilage of an animal is able by bond 500 g case of khakhra (*Bute amonosprema*) and 500 g berry of guvar (Delonixelata).

Source: https://www.researchgate.net/publication/282818619_Use_of_ITK_in_Plant Protection

Management of paddy insect pests through *Vitex negundo* (Nirgundi): Farmers domain the paddy land with brooms made-up up of branches of *Vitex negundo,* which are accustomed to act as an insecticide and enhance advance of paddy.

1. Citrus plants: 4 kg tobacco leaves and twigs are aloft in 40 liters of water for 40 min. Afterwards cooling, 1 Kg soap acquaintance is alloyed and is adulterated 7-8 times and sprayed to get superior control of aphids and white flies in citrus plants; Also the practices of cloudburst kerosene or petrol in holes and blocking holes with mud or cow dung are as well accustomed in citrus plant;
2. Rice hopper: Rice seedlings uphill from berry advised with abstract of neem leave molecules are aggressive to blade hopper; A Band-Aid able from neem blade adhering in water (10 kg: 2 liter) is able to ascendancy blade binder in rice;
3. Mango trees: for blockage of bane of shoot auger in mango tree, accustomed alkali is alloyed with clay abreast the collar arena of tree; In case of insect holes fabricated by shoot auger and case eaters in mango, jiggery is placed in the holes to allure added predators to augment on the insect present in the holes.
4. Chilli: 5. 1 liter abstract of according abundance of ashamed blooming chilies and garlic alloyed with 200 liters of water is able to ascendancy aphids and jassids;
5. Sucking pests: Filtrate able from a Band-Aid of 1 kg *Calotropis* leaves and 5 liters of water is able to superiority on leave sucking pests;
6. Axis borer: An Extract of tobacco decay with 250 g of soap Band-Aid in 200 liters of water as spray superiority to axis borer.
7. Insect killer: A Band-Aid able from 100 g tulsi (*Ocimum sp.)* leaves in 1 liter of water with 1 ml soap Band-Aid can be acclimated for able ascendancy of aphid, army worm, red affection bug, mosquito etc. is acclimated to superiority to protect from semilooper, mites, blooming blade hopper, aphid etc.

Control of Stored Grain Pests

1. Garlic leaves are acclimated for safe accumulator of aliment grains;
2. Leaves of *Vitex negundo* (Nirgundi) are congenital in any beating seeds for continued time preservation;
3. There is an accustomed conveyance of autumn aliment grains like aureate and rice use of neem leaves to anticipate accumulator annoyance damage;

4. Milled chickpea, blooming gram and added pulses are stored afterwards thoroughly advised with alacrity oil;
5. Green gram can be kept charge less from any annoyance bane by alleviative with 1% neem blade powder\;
6. Seed alloyed with *Acoruscalamus* (baje) crumb in the adjustment of 10 kg: 1 kg would advice in attention the berry charge less from stored pests for continued time.

Conventional saver structures: It has been found that use adequate accumulator structures such as mud bins, rock bins and bamboo bins for storage are popularly used before storage, they acclimated to antisepticise the grains with smoke of cow gung block and neem leaves, the majority of farmer's large quantity of grains and meat abreast the kitchen extent the smoke of fire firewood penetrates. They as well use leaves of neem or tulsi on the accumulator anatomy to accumulate charge less from insect-pest infestation. Use rat allurement supposed to be called'gurung'. The allurement is fabricated of tauk (thin bamboo) with continued internodes. Garo tribes of Meghalaya use particle collector structures fabricated up of beard grass, bamboo and lath poles. An adequate adjustment of rat affirmation collector locally alleged nahu (granary) can be apparent at the one bend of angel.

Conclusion

The literatures accumulated through the world, this is established fact that the plant disorders with specific causes along with symptoms were being treated with the materials available indigenously and effective plant protection has been provided to crops, plants, trees since ancient period and those were originated from indigenous traditional knowledge (ITKs)

Some pertinent short questions

1. Write down the role of ITKs in controlling diseases;
2. Write down the role of ITKs in controlling pests;
3. What is trapping bake-apple flies and which crop the pests are found?
4. What is the role of ITKs in controlling store grain pests?
5. Write down the use of lantana ashes in controlling the pests;

11

Crop Voyage In India and The World

Introduction

The literary meaning of voyage is a long journey by water or air to a distance place. The crop voyage means the introduction of different crops from outside and movement of the crops originated in India to abroad. India is one of the mega agro-biodiversity countries of the world. Many plant species have been domesticated here. The agro-biodiversity of the indigenously domesticated higher plant species was further enriched with introduction of exotic crops since ancient times, as reflected by the presence of American cereals such as maize and amaranths, African cereals such as pear millet and sorghum, West Asian legumes such as field pea, grass pea and chickpea, and Chinese *proso* millet and buckwheat in the archeological remains and/or in ancient Indian agriculture writings. Most of these crops have acclimatized and adapted to diverse ecological conditions offered by the Indian Sub-continent to such an extent that many appear natural to the land, and have thrown enormous genetic diversity, making India either a secondary center of diversity for crops such as pearl millet, sorghum, maize, amaranths, cowpea, niger, safflower, muskmelon, pumpkin, chilies and others or regional center of diversity for crops such as barley, grain amaranth, buckwheat, pearl millet, foxtail millet, chickpea, bottle gourd and others. All details of crop voyage has already been discussed in the relevant topic of discussion (Chapter 8) on "Journey of agriculture and its development from past to modern era."

The hypotheses of the crop voyage

The hypotheses about ancient introductions are based more on presumptions of evolutionary biologists and researchers rather than on hard data belonging to evolution of agriculture, societies, economics, and trade. Systematic and scientific investigations in world crops such as rice, wheat, maize and others have revealed that much information in this regard can be deduced from archaeo-botanical, archeo-zoological and literary records.

The ancestry of agricultural in eastern Asia date to Neolithic times, 7000 to 9000 years ago, with rice agricultural about 4000 BCE. Agronomical origins in the Indus basin took action about the aforementioned time. In North China and Manchuria, an acculturation was accustomed about 2000 BCE and by the Zhou dynasty, 1000 BCE, there is affirmation of canals and all-encompassing irrigation. The writings of Confucius (551–470 BCE) acknowledgment 44 aliment plants including agronomical crops such as peach, plum, Japanese apricot (Prunusmume), jujube, chestnut, mulberry, quince, Chinese cabbage, canteen gourds, and assorted melons. First year agronomical manuals alarm accelerated assembly of crops, pretreatment of seed, anhydrous rice with broadcast water, backbone cultivation, pot irrigation, crop scheduling, composting, and adamant tools. Books on agricultural and agricultural are produced amid 221 and 550 CE. Accessory agricultural became anchored in the ability of China and advance throughout Asia with the development of rural retreats and burgher breadth of the emperors. Flower agricultural became one of the seven arts and affected abstruse importance. Barter of crops and technology through barter and conquests amid East and West has an age-old and continuing attitude as is apparent by cottony strands on Egyptian and casket in 1000 BCE. The biblical exchange in spices complex both sea routes via India and the Mideast as able-bodied as overland routes (the Cottony Road) through Persia, and it is by this avenue that the peach, mulberry, and citrus accomplished the West. European incursions in the Americas in the 'Age of Exploration' added complete East-West contacts and fresh crops such as hot pepper, maize, candied potato, potato, and peanut become important crops in China. Recent Asian innovations in agronomical and agronomical technology cover advances in artificial ability through the development of activity extenuative greenhouses, grafting technology for vegetable production, and the development of amalgam rice. Asia is one of the continents in which acculturation first developed and breadth bodies abstruse to reside by cultivating plants. In fact, added than bisected of the world's comestible aliment crops originated in Asia. Admitting the actuality that Asian Agronomical Technology has and continues to accept an astronomic aspect on the world, its agronomical history has seemed perplexing to the West and ignored. The huge self-disciplined alarmed Asia is generally lumped with Europe and the Mid-East as Eurasia but there is no absolute jump to this sufficient land area. However, Eastern Asia, which is accepted to the West as the Orient is ecologically afar from the Western allotment of the self-denying by dreadful limitations: chastening and massive large quantity of ranges. Although associated amid East and West is actual ancient, it was abundantly unusual through exchange by sea via India and Arabia and overland through the road-ways. Western acclimatization of archeological analysis in the Mideast has resulted befitting nation in that span as the "Garden of Eden" from which all account be originated. This may be an inaccurate account of history. Of the seven crop centers for the agents of crop plants delineated by the rich Russian

Bio-geographer N.I. Vavilov, three are Asian. Center I, is the Tropical South Asiatic Center which includes India and South Eastern Asia; Center II, is the East Asiatic Center, which includes locations of China and Japan; and Center III, the Southwestern Asiatic Center, referred to now as Axial Asia.

In view of the worlds acceptable agricultural and horticultural crops that accepted to be advised as an allowance from Asia. In contrast, as an aftereffect of algidity during the Ice Age, the arctic areas of the locations such as Siberia, Arctic Europe, Canada, and the Continental United States accommodated few built-in type and accept contributed almost few grain crops; affirmation grasses, sunflower, and some little one fruits such as strawberries and vaccines (blueberry, cranberry, and lingo berry) are examples. As a consequence, Asian germ-plasm assets accepted to be advised for approaching crop improvement. Admitting the alternate exchange of affluent crops amid Asia, Europe, Africa, Oceania, and the Americas that has taken residence over the centuries, not all were included, opening up the possibility of new exchanges.

History of civilization and agricultural technology

The acceleration of the age-old Chinese dynasties begins in the fertile, silt-enriched land coverage under Yellow River basin. Political institutions in the year about 2000 BCE revolved about the aristocratic city-limits with affirmation of writing, bronze, acclimatized carts, and divination by basic and tortoise shells. The age-old kingdoms of Xia (21–16 year BCE), Shang (16–11 year BCE), Zhou (1066–221 BCE), and clashing states (Qin, Zhao, Wei, Han, Yan, Qi, Song, Chu, Lu) were eventually unified by 221 BCE in the Qin totalitarianism (221-207 BCE). Any of above accessible works by the emperor resulted in the planning of an arrangement of waterfront glowing from the capital, irrigation canals, arresting walls, and gigantic palaces and tombs. Abundant Wall (221–207 BCE) was completed and afterwards profusely transmitted as guidance in adjoining the clashing freezing type of tribes. Korea was annexed followed by acquisition of the accomplished of Axial Asia during the Han despotism (206 BCE–220 CE), above industries included cottony and unyielding with the intensification of exchange to the West. An abrupt outline of China history (Encyclopedia Britannica, 1970) emphasizing complex advances is to actual by repression that is given below:

Historical the agent of the subcontinent is based on an age-old blow of a land area with Asia to consequences to the Himalayas, familiar amid China and India. The name India (a Greek word) derives from the Sanscrit Sindhu (Persian Hindu), and accredit to the 1000 mile river that bisected this enormous, populous, polyglot subcontinent. The altitude varies from the wet monsoons to deserts, from the snows of the Himalayas to the hot jungles of the south. India is not now, and never was, an individual nation. Some of the native invaders were alleged Aryans or power lover, who acquire from peoples from the shores of the Caspian Sea, whose western annex busy Europe. Local populations are supposed as Dravidian. India entered Western watchfulness afterwards Alexander's aggression of Western India (327–325 BCE). Original agnostic religious behavior (Veda) developed into Hinduism, accumulation wealthy Gods with a claimed affiliation amid the adherent and the divinity. Buddhism, founded by Gautama Siddhartha (563–483 BCE), an attachment of Hinduism, would advance to axial and eastern Asia span and it became an increasing religion. Society developed in a familial degree system: (i) *Brahman* (religious specialists), (ii) *Ksatriya*, (warriors and political leaders), (iii) *Vaisy* (farmers, craftsmen, and merchants) and (iv) *Sudra* (servants to the added three); the arrangement still persists admitting activist attempts to eliminate it.

Influences on advancement of Indian agriculture

India was abundantly influenced by two incursions from the 7th to 13th centuries. The eastern ambits of Islamic warriors that started a few centuries afterwards while the Arabian Philosopher and Prophet Mohammed (570–632) and the invasions of the Mongols (Moguls), allotment of the conquest of Cgenghis Khan (1115–1227). The two incursions were to adhere if Timur Lenk (Tamerlane), a Mongol invader, adopted Islam and raided Delhi in the year 1398. As an aftereffect of Islamic incursions, the Indian subcontinent was break into an alternation of Muslim and Hindu kingdoms and principalities that existed in about its prosperity. These were about plucked by Britain in the 19th century through colonial domination. British India was disconnected amid the Muslims (Pakistan and the noncontiguous East Pakistan) and the secular state of majority Hindus (India). The countries of the subcontinent now abide (from West to East) of Afghanistan, Pakistan, India, and Bangladesh (formerly East Pakistan) with Nepal and Bhutan huddled in the Himalayas. Indian agricultural in the Indus basin is advised one of the areas in which agricultural operations were apparent. Agronomic lineage arises in the 6th millennium BCE. The charcoal of age-old cities allusive those of Sumer, Babylonia, and Egypt are begin in two sites, Mohaenjo-daro and Harappa that dated about 3000 BCE. Evidently, they cover pottery, dice, chessmen, coins, seals that buck profound inscriptions, carvings, weapons, implements, and ornaments, and a huge granary on a brick with an air-drying system. Accoutrement was made-up of stone, copper, and bronze were found. The native peoples of southern India were agriculturists who tilled the soil, uphill the crops of barley, grain crops (rice was an afterwards introduction), and affection and calm horned animals, poultry, and apparently elephants. The afterwards Hindu prohibitions linked up meat advance an absence they say that due to an access in inhabitants. Hindu prohibitions attached meat accepted larboard plentiful amongst the Indian subcontinent vegetarian but dairy articles are used. From these beginnings, agriculture in India continued to rice, pulses, millet, vegetables, and fruits. Dams for irrigation were completed as indigenous as in the 1st century. An Indian cooking developed that is affluent in spices such as curry, ginger, cloves, and cinnamon, comestible resources approved by the Europeans. One of the indigenous crops was cotton, accepted to the ancients as tree-grown wool. Affluent of names for fabrics are acquired from India (muslin derives from affection weavers of Mosul, calico, from Calicut). The rugs of India are still award-winning. Exchange in Indian accessories was intensely basic in the age-old world; Indian cheetahs, tigers, elephants assisted in the blood-soaked aggressive games. The great quantity of India became legendary, extensive its height in the 14th century. India was made-up considerable contributions to medicine. Two systems were developed, (i) one based on Hinduism (Ayrveda) and other (ii) the added Moslem (Unani). In the Ayrveda system, drugs are mainly from rhizome agent (herbal medicine), while the Unani system, developed later, and as well incorporate drugs acquired from animals.

Liaison and relationship between East and West

Contacts amongst East and West accepted a connected but haphazard history. Exchange of populations appears to be accepted apparent the elderly period; in actuality the indigenous society got trouble due to violence of Cro-Magnon bodies into Europe acquire from Asia. Subsequently contacts in the middle of East and West accepted two types, (i) an almost peaceful one generated by trade and (ii) another an advancing one apprenticed by fight and conquest. Confirmation of cottony strands in an Egyptian casket in the 10th BCE suggests actual indigenous exchanges of accessories amid China and Egypt it appears that it came via overland routes that included Persia. This was apperceived of the age-old smell exchange amid the Far East and Egypt from the biblical adventure of Joseph in Genesis and again allusion to spices such as biscuit and cassia. These admired spices, which originated in the Maluccas Islands (presently portion of Indonesia), gifted the Near East overland or by sea via India.

Contrarians

However, in following the above criteria, the difficulties listed below were likely to be encountered.

1. Limited archeological evidences in crops propagated by vegetative means (leafy vegetables, root and tuber crops);
2. Lack of knowledge of oriental languages, such as Sanskrit, Prakrits, Pali, Dravidian, and other ancient languages;

3. Difficulties in identification and verification of Latin synonyms of plant species;
4. Poor documentation and inadequate information regarding minor crops of marginal significance;
5. Difficulties in distinguishing the ancient introductions from those from medieval period because of lack of indigenous information and pro-west attitude of historians. In crops of African origin, access to the Indian Subcontinent, both through land and sea-route, is possible; and
6. It is difficult to distinguish between the two routes and in some crops, it might have occurred through both. Phylo-genetic studies including use of molecular markers may be needed to decide on the nearest dates and relationships with existing genetic diversity.

Crop voyage and domestication

The ancient evidences pronounced introduction and domestication of different crops in Indian subcontinent and voyage of crops on India origin to other corners of the world.

Kharif crops: *Kharif* crops are usually sown with the beginning of the first rains in July, during the south-west monsoon season. Major *Kharif* crops of India include Millets (Bajra and Jowar), Cotton, Soybean, Sugarcane, Turmeric, Paddy (Rice), Maize, Moong (Pulses), Groundnut, Red Chillies, and others.

Rabi crops**:** *Rabi* cropa are the spring harvest or winter crops are sown in October last and harvested in March April every year.

Major Rabi crops include Wheat, Barley, Mustard, Sesame, Peas and others.

Zaid crops: These crops are grown in some parts of country during March to June. Prominent examples are Muskmelon, Watermelon, Vegetables of *Cucurbitacae* family such as bitter gourd, pumpkin, ridged gourd and others.

Major crops grown in India

Rice: Rice is predominantly a *kharif* crop covers one third of total cultivated area of India. It provides food to more than half of the Indian population. Rice is produced in almost all states. Top three producer states are West Bengal, Punjab and Uttar Pradesh. Other rice growing states include Tamilnadu, Andhra Pradesh, Bihar, Jharkhand, Uttarakhand, Chhattisgarh, Odisha, Uttar Pradesh, Karnataka, Assam and Maharashtra. It is also grown in Haryana, Madhya Pradesh, Kerala, Gujarat and Kashmir Valley.

Wheat: Wheat is the second most important crop of India after Rice. It's a *rabi* Crop. It is the staple food in north and north western India. It's a winter crop and needs low temperature. Ideal temperature for wheat cultivation is between 10-15°C at the time of sowing and 21-26°C at the time of harvesting. Wheat thrives well in less than 100 cm and more than 75 cm rainfall. The most suitable soil for cultivation of wheat is well drained fertile loamy soil and clayey soil. Plain areas are most suitable. The wheat crop is highly mechanization oriented and may need less labour. Top three states producing Wheat are Uttar Pradesh, Punjab and Haryana.

Coarse Cereals / Millets: Coarse Cereals and Millets are the short duration warm weather (*kharif*) crops used both as food and fodder. Important millets are *Jawar, Bajra, Ragi etc.* The areas under these crops have fallen drastically in recent years in India. The coarse cereals and millets are grown in areas with high temperature and are called dryland crops because can be grown in areas with 50-100 cm rainfall. The coarse cereal crops are less sensitive to soil deficiencies. They can be grown in inferior alluvial or loamy soil. Top three states with maximum production of total coarse cereals are Maharashtra, Karnataka, and Rajasthan.

Pulse crops: Most pulses are leguminous crops and provide proteins to the vegetarian population. Major pulses of India include Gram, T*ur or arhar* (Pigeon Pea or Red Gram), *urd* (black gram), *mung* (green gram), *masur* (lentil), *kulthi* (horse gram), *matar* (peas) etc. But among these above mentioned varieties only gram and *tur or arhar* are more important pulses.

Sugarcane: Sugarcane belongs to bamboo family of plants and is indigenous to South Asia. In India, it is one of the most important *kharif* crops.

Cotton: Cotton is the most important fibre crop and cotton seed is used as a vegetable oil and a part of fodder for milch cattle for better milk production. Cotton is a kharif crop and grows in tropical and subtropical areas. Cotton requires modest rainfall, it is one of the predominant rainfed crops. Cotton requires uniformly high temperature (21°C to 30°C). It grows in areas having at least 210 frost free days in a year.

Optimum soil for cotton is the Black soils of Deccan and Malwa plateau. Also grows well in alluvial soils of the Satluj-Ganga plain and red and laterite soils of the peninsular region. Cotton growing is known as less mechanized farming in India so needs cheap labour. Main cotton producing states are Gujarat, Maharashtra and Andhra Pradesh.

Groundnut: Groundnut is most important oil seeds of India. Grown as both as kharif and Rabi crop but 90-95% of the total area is devoted to kharif crop. Groundnut thrives best in the tropical climate and requires 20°C to 30°C temperature. 50-75 cm rainfall is favourable for groundnut cultivation. Groundnut is highly susceptible to frost, drought, continuous rain and stagnant water. It needs dry winder at the time of ripening. Well drained light sandy loams, red, yellow and black soils are well suited for its cultivation. Groundnut accounts for half of the major oilseeds produced in India. India is the second largest producer of groundnut (After China). Top three states producing ground nut are Gujarat, Andhra Pradesh and Tamil Nadu.

Tea: India is the largest producer and consumer of black tea in the world. Tea is grown in 16 states in India. Assam, West Bengal, Tamil Nadu, and Kerala account for about 95 per cent of total tea production. More information about tea production in India can be obtained from this link.

Coffee: Coffee needs hot and humid climate with temperature varying between 15°C and 28°C. It is generally grown under shady trees. Strong sun shine, high temperature above 30°C, frost and snowfall are harmful for coffee cultivation. Dry weather is necessary at the time of ripening of berries. Rainfall between 150 to 250 cm is favourable for coffee cultivation. Well drained, rich friable loamy soil with humus and minerals are ideal for coffee cultivation. Coffee also needs cheap and skilled labour. Major coffee producing states of India are Karnataka, Kerala and Tamil Nadu. More information about Coffee production in India can be obtained from this link.

Factors affecting crop diversification

With the advent of modern agricultural technology, especially during the period of adoption of advanced technologies in the late sixties and early seventies, there is a continuous surge for diversified agriculture in terms of crops, primarily on economic considerations. The crop pattern changes, crop diversification, however, are the outcome of the interactive effect of many factors which can be broadly categorized into the following five groups:

a) Resource related factors covering irrigation, rainfall and soil fertility.

b) Technology related factors covering not only seed, fertilizer, and water technologies but also those related to marketing, storage and processing.

c) Household related factors covering food and fodder self-sufficiency requirement as well as investment capacity.

d) Price related factors covering output and input prices as well as trade policies and other economic policies that affect these prices either directly or indirectly.

e) Institutional and infrastructure related factors covering farm size and tenancy arrangements, research, extension and marketing systems and government regulatory policies.

Although the factors that influence the area allocation decision of farmers are all important, they obviously differ in terms of the relative importance both across farm groups and resource regions. Factors such as food and fodder self-sufficiency, farm size, and investment constraints are important in influencing the area allocation pattern among smaller farms, larger farmers with an ability to circumvent resources constraints usually go more by economic considerations based on relative crop prices than by other non-economic considerations. Similarly, economic factors play a relatively stronger role in influencing the crop pattern in areas with a better irrigation and infrastructure potential. In such areas, commercialization and market networks co-evolve to make the farmers more dynamic and highly responsive to economic impulses. A notable is the change in the relative importance of these factors over time from generalized perspective to Indian agriculture that is increasingly getting influenced more and more by economic factors. This need not be surprising because irrigation expansion, infrastructure development, penetration of rural markets, development and spread of short duration and drought resistant crop technologies have all contributed to minimizing the role of non-economic factors in crop choice of even small farmers. The reform initiatives undertaken in the context of the ongoing agricultural liberalization and globalization policies are also going to further strengthen the role of price related economic incentives in determining crop composition both at the micro and macro levels. Obviously, such a changing economic environment will also ensure that government price and trade policies will become still more powerful instruments for directing area allocation decisions of farmers, aligning thereby the crop pattern changes in line with the changing demand-supply conditions. In a condition where agricultural growth results more from productivity improvement than from area expansion, the increasing role that price related economic incentives play in crop choice can also pave the way for the next stage of agricultural evolution where growth originates more and more from value-added production.

Consequences of changes in crop pattern

Turning now to the socio-economic and environmental consequences of crop pattern changes, the modern agro-technologies have fomented, amongst other things, an increasing tendency towards crop specialization and commercialization of agriculture. These developments have positive effects on land/labour productivity and net farm income, also endangered a number of undesirable side effects like reduced farm employment and economic imbalances. Although the expansion of commercialized agriculture has fomented new sets of rural non-farm activities and strengthened the rural-urban growth linkages, it has also weakened the traditional inter-sectoral linkages between the crop and livestock sectors. Besides, crop pattern changes also lead to serious environmental consequences that take such forms as groundwater depletion, soil fertility loss and water-logging and salinity. All these can reduce the productive capacity and growth potential of agriculture over the long-term.

Success story in crop diversification

It is clear that most of the area shifts that occurred during the three decades between 1966-67 and 1996-97 are from coarse cereals to oilseeds. Three notable aspects of this area shift towards oilseeds can be noted. First, most of these area shifts have occurred particularly during the decade ending 1996-97. A protective trade environment, favourable price policy and the connecting of the Technology Mission on Oilseeds (TMO) during this period have all enhanced the comparative advantage of oilseeds. The area share of oilseeds that increased just by a percentage each during the two decades before 1986-87 has risen suddenly by 43 percent during the decade ending 1996-97. Second, the area shift during this period comes mainly from barley and pulses other than pigeon pea. Since most of those crops losing their area share are usually grown under rainfed conditions where oilseeds can also be grown, the area shift can be said to involve mostly rainfed areas, although comparative advantage and crop rotation considerations often favour oilseeds even in groundwater irrigated areas. Thirdly, there is also a significant area shift within oilseed crops. The area shares of rapeseed and mustard, sunflower and soybean are increasing steadily, those of sesamum, linseed and nigerseed are declining gradually. Thus, the area shift has favoured only a sub-sector within the oilseed sector partly because of constant changes in the comparative advantage of different oilseeds and partly because of the impact of changing consumers' preferences on the relative demand of oilseeds. Area shifts and crop pattern changes can lead either to crop specialization or to crop diversification. The area share of food grains increased during 1967-76 due partly to their yield advantages created by irrigation expansion and introduction of high yielding fertilizer and irrigation responsive cereal crops with modern agro-technologies and partly to government policies pursued to encourage food production and eliminate food imports. As a result, there was a tendency towards cereal centered specialization. But, later when increased productivity of food grains, especially cereals, made it possible to allocate more area to other crops such as oilseeds with a severe supply shortage, the specialization tendency witnessed earlier has given room for overall crop diversification. Even within such an overall crop diversification, it is also possible to see specialization tendencies within each crop group. Such a specialization has become possible because within the achievement of self-sufficiency, mainly through an expanding production of wheat and rice, the food implication of coarse cereals has declined almost coinciding with their declining demand prompted by an increasingly incurring income level. Within oilseeds, even though groundnut still has a dominant area share, the growth of area under rapeseed and mustard, sunflower and soybean indicates an ongoing structural change leading to diversification within the oilseed sector. The issue of crop diversification, although considered in area terms, assumes significance in view of its effects on the supply-demand balance of main crops and crop groups. Rice and wheat-centered specialization, for instance, indicates an increase in the supply of wheat and rice but a reduced supply of coarse cereals. Since the demand for coarse cereals is declining and that for wheat and rice is increasing due to changes in the income pattern, the changes in their supply are actually necessary to achieve the required demand-supply balance. A similar line of argument can also be extended to other crops where in addition to domestic demand, international demand and supply also assume significance. The food grains category has evidenced as an increasing tendency towards specialization with an unbalanced area composition, whereas the non-food grains category has shown an increasing tendency to diversify with an improved balance in their area composition. But, during 1976-97 the non-food grains category has also caught up with the specialization tendency and area imbalance that characterized food grains throughout the entire period. The reason for the specialization tendency within food grains is rather obvious in view of the declining area share of coarse cereals and the increasing area share of rice and wheat. The changing area share of crops can also explain the specialization tendencies observed among the non-food grains. As the area shares of oilseeds and other commercial crops have increased especially during the period characterized by a prices and market-responsive agriculture inevitable in the aftermath of achieving food self-sufficiency. It is important to note, however, that oilseeds are more diversified with an enhanced balance in their area composition. Within food grains, on the other hand, while cereals tend towards specialization with an increasing imbalance in their area composition, pulses evince a tendency to diversify with a declining imbalance in their area composition. The latter result is mainly due to the fact that the area shares of the dominant pulses such as gram have declined to give room for other pulses such as pigeon pea. But it should be noted that diversification and area balance achieved by pulses have occurred only with a reduced area share. Among the non-food grain crops, spices are notable not only for their increasing tendency towards diversification but also for a declining imbalance in their area composition. In contrast, fibres, fruits and vegetables, other field crops and plantation crops have all evinced a specialization tendency especially after 1986-87.

Production and productivity effects

The evaluation of crop pattern changes and their implications for crop diversification and area composition cannot be completed without considering their output and productivity effects. The production and productivity effects of area shifts can be evaluated by considering both the growth rates of output, area, and yield of various crop groups as well as the crop group specific patterns evident in the relative contributions of area and yield to overall output growth. The growth rates and sources of growth for the major crop groups at the all India level for the three periods, i.e., 1966-76, 1976-86 and 1987-97. The shift of area from cereals (especially coarse cereals) has not led to any decline in their output thanks to an impressive performance of their yield levels. The productivity improvement in cereals has been such that it compensates even for the area loss of cereals occurred during 1986-97. Notably, in the case of coarse cereals that have suffered a heavy area loss ever since the year 1976, yield improvement has been substantial especially in the last period. As a result their output growth, which was only marginal during 1976-86, has jumped to 1.14 percent during 1986-97. Pulses, on the other hand, even though area growth has increased from 0.21 to 0.32 percent between the last two periods, their output growth turns out to be negative due to a fall in their yield levels. Food grains showing a steady rise in their output growth have a story similar to that of cereals as the contributions of their yield growth is increasing; whereas, those of their area growth are declining even to become negative in the recent period. Turning to the growth rates and growth sources of oilseeds, it is useful to distinguish the trend observed for the three oilseeds, i.e., rapeseed and mustard, sunflower and soybean from that observed for the oilseeds as a whole. This is because of the fact that the area shares of these three oilseeds have grown tremendously since 1976. Their combined area share of 6.85 percent observed at present is higher than the combined share of both groundnut and sesamum, which dominated the oilseed sector in the pre-Modern Technology era due to their faster expansion. The output growth of nine oilseeds has been impressive as compared to that of other crop groups; it is not as impressive as the combined output growth of the three fast growing oilseeds. However, there is a marked difference in the sources of output growth between the nine oilseeds and its sub-group. Although the contribution of area growth dominates that of yield growth in both cases, the contributions of area growth are far higher in the case of the three oilseeds as compared to that in the case of nine oilseeds. It should be noted that it is the higher area growth observed in the case of the three oilseeds, which has also compensated for the declining area share of some of the traditional oilseeds such as sesamum, linseed and niger seed. Since the yield growth of the three oilseeds is also higher than that obtained for the nine oilseeds, the productivity improvement in the case of former is also relatively higher than that of the traditional oilseeds. There is also a notable shift in the sources of output growth among the commercial crops (i.e., crops other than cereals, pulses, and oilseeds) as yield growth that dominated over the area growth during 1966-1986 becomes dominated in turn by a area growth during 1986-97. The implications are clear that area shifts from crops need not be a problem as long as their productivity levels are increasing faster to compensate for their declining area share. This is what has happened especially in the case of coarse cereals that have been a net donor of area to oilseeds and other commercial crops. On the other hand, since the area shifts were from low-value coarse cereals to high-value oilseeds and since such shifts were accompanied by yield improvements within the oilseed sector, the area shifts have not only increased the overall output of edible oils but also contributed to an enhancement of the income level of farmers. In this sense, the area shifts have actually contributed to broaden the foundation of food security. Poverty eradication is one of the major objectives of plan development. The magnitude of the problem is quite staggering. The incidence of poverty declined from 54.9 percent in 1973-74 to 36 percent in 1993-94. The absolute number of population did not decline much throughout this period of 20 years. There were 321 million poor in 1973-74 and 320 million in 1993-94. In the rural areas, the corresponding numbers were 261 million and 244 million. The land resources are limited and the average size of holding is only 1.57 ha for the census year of 1990-91. The main determinants of poverty are: i) lack of income and purchasing power contributable to lack of productive employment; ii) the continuous increase in price of food, specially food grains which account for 70-80 percent of the consumption; and iii) inadequacy of social infrastructure, affecting the quality of life of the people and their employability. National agenda of the present government has given top priority to doubling food production in the next 10 years. The food includes food grains (rice, wheat, coarse cereals, pulse crops), oilseeds, sugar (sugar cane), fruits and vegetables, meat, milk, and fish. Action Plan envisages a detailed strategy and specific problems of productivity to substantially increase the supply of various food items in such a way that the demand for such items for the entire population is comfortably met and some exportable surplus also becomes available. The development strategy to be perused in the medium term has been consciously interwoven with the country's food security concern. It is known fact that there is little scope for further expansion of the net sown area (142 million ha) and that land scarcity will become an acute feature of the rural economy. Water is a precious national asset and there are several concerns regarding water resources in the country. Therefore, a judicious use of land and water resources will have to be the central theme for sustainability of agricultural growth. There has been a growing concern in recent years about the deteriorating conditions of soil health and water resources due to improper management and pollution. Deterioration in land and water resources has been in the form of land degradation, water logging and decline in water table. There is a greater need to have an integrated approach in the management of plant nutrients, chemicals and taking effective measures to deal with the overall pollution problems. There are several possible technologies and alternatives to reduce the use of chemicals in agriculture. These alternatives are not perfect substitutes to chemicals but adoption of these can substantially reduce the adverse impact on environment. Proper land and water management policies would reduce environmental degradation. Community and village institutions should be encouraged to participate in protecting natural resources from degradation and the research programmes for regeneration of land and water resources are highly on demand to be strengthened.

Conclusion

Crop voyage, from these beginnings, agriculture in India continued to rice, pulses, millet, vegetables, and fruits. Dams/reservoirs for irrigation were complete as early as the 1st century. An Indian food developed that is prosperous in spices such as curry, ginger, cloves, and cinnamon, edible and palatable food treasures approved by the Europeans. One of the early as crops was cotton, accepted to the ancients as tree-grown wool. Affluent of names for fabrics are acquired from India (muslin derives from affection weavers of Mosul, calico, from Calicut). The rugs and carpets of India are still prized. Trade in Indian trappings was acutely basic in the age-old world; Indian cheetahs, tigers, elephants assisted in the blood-soaked Roman aggressive games. The great quantity of India became legendary, extensive its height in the 14th century. India fabricated contributions to medicines from plant orogin. Two systems were developed, one based on Hinduism (Ayrveda) and the added Moslem (Unani). In the Ayrveda system, drugs are mainly from rhizome agent (herbal medicine), while the Unani system, developed later, as well incorporate drugs acquired from animals. Unani medicines are natural process that control disease with symptoms and reactions to the the body of human/animal. It employs the humoral theory which presupposes the presence of four akhlaat (humours) in the body, dam (blood), balgham (phlegm), safra (yellow bile) and sauda (black bile).

Some pertinent short questions

1. What is crop voyage in agricultural heritage?
2. Effect of crop voyage on production and productivity of crops;
3. What is natural resource management in crop cultivation?
4. Write down hypotheses of crop voyage in agricultural heritage;
5. Role of crop voyage in food and nutritional security;

12

Medicinal Plants and Thier Importance in Agricultural Heritage

Introduction

Medicinal plants are considered as rich ingredients resources which may be utilized in medicines development pharmacopoeia, non-pharmacopoeia or synthetic industry. Apart from any region, these plants play a critical role within the development of human cultures around the whole world. Plant is a crucial source of drugs and plays a key role in world health. Medicinal herbs or plants are known to be a critical potential source of therapeutics or curative aids. The utilization of medicinal plants has attained a commanding role in health system all over the place the planet. This involves the utilization of medicinal plants not just for the treatment of diseases but also as potential material for maintaining healthiness and conditions. Many countries within the world, that is, two-third of the world's population depends on herbal medicines for primary health care. The explanations for this are often due to their better cultural acceptability, better compatibility and adaptableness with the physical body poses lesser side effects. Evidently, records showed that most of the used drugs contain plant extracts. Some are containing active ingredients (bioactive components or substances) obtained from plants. Through recent researches, it revealed that plant-derived drugs were mostly discovered from the study of curative, therapeutic, traditional cures and above all the folks knowledge of indigenous people and a few of those claims and believe of individuals are irreplaceable despite the recent advancement in science and technology. A number of the drugs believed to be obtained from plants are aspirin, atropine, artimesinin, colchicine, digoxin, ephedrine, morphine, physostigmine, pilocarpine, quinine, guanidine, reserving, taxol, tubocurarine, vincristine and vinblastine. The importance of medicinal plants can't be taken lightly; therefore, it'll be of significance to look at the characteristics and roles of photochemical in a number of the medicinal plants commonly utilized throughout the world.

Definition of medicinal plants

Medicinal plants have a considerable number of definitions to propose the term 'medicinal plant'. Consistent with the World Health Organization, "a medicinal plant is any plant which, in one or more of its organs, contains substances which will be used for therapeutic purposes, or which are precursors for chemo-pharmaceutical semi synthesis". This definition allows for a distinction between the already known medicinal plants whose therapeutic properties or character as a precursor of certain molecules are established scientifically, and other plants utilized in traditional medicines and regarded medicinal, but which haven't yet been subjected to thorough scientific study.

Plants for human health

Today's health care systems rely largely on materials available from natural sources. Much of the world's population depends on traditional medicine to satisfy daily health requirements, especially within developing countries. Use of plant-based remedies is additionally widespread in many industrialized countries and various pharmaceuticals are supported or derived from plant compounds. Similarly, cosmetics and other household products may contain plants of medicinal or therapeutic value. The pharmaceutical industry is both large and highly successful. Sales of plant derived drugs are expected to succeed worldwide in 2002. At the present about 50% of the entire plant-derived drug sales come from synthetic entities, while the remaining 50% come from herbal remedies. Although the latter have greater volumes of consumption, the relatively low volumes of synthetic entities, which are mostly prescription products, are quite compensated by their higher prices. Those plant drugs, which mostly treat serious medical bills, include atropine, dioxin, morphine, paclitaxel, pilocarpine, reserpine, scopolamine, topotecan and vincristine, among many others. Several of the compounds have outlived their usefulness in light of higher alternatives, however, and are exhibiting decline in sales.

South Asia is home to several rich, traditional systems of drugs. Ayurvedic system dates back to 5000 B.C. along side the Unani, Siddha and Tibetan systems, these traditional system of medicines (TSMs) remain important source of everyday health and livelihood for tens of many people. Himalayan sage-scholars of traditional medicine have said "*Nanaushadhi Bhootam Jagat Kinchit*" *i.e.* `there is not any plant within the world, which doesn't have medicinal properties.' The traditional scholars are estimated to understand the medicinal properties of many species of plants. It's therefore, no exaggeration to mention that the uses of plants for human health are probably as old as citizenry concepts amongst themselves. Even so, the recent dramatic increase in sales of herbal products in global markets underscores the growing popularity of herbal therapies. This has created new opportunities for the countries, their largely impoverished populace and traditional herbal industry; it also poses unprecedented threats to the very resources on which the industry depends besides creating socioeconomic imbalances and erosion of spiritual and cultural heritage and knowledge systems. Medicinal and aromatic plants (MAPs), including trees, shrubs, grasses and vines, are a central resource for these traditional health systems, furthermore as for pharmaceutical (or allopathic) medicines. There are moderately 8,000 plant species in South Asia with known medicinal uses. Medicinal plants are accessible, affordable and culturally appropriate sources of primary health look after a pretty 80% of Asian population. Poor and marginalized, who cannot afford or access formal health care systems, are especially hooked in to these culturally familiar, technically simple, financially affordable and usually effective traditional medicines. As such, there's widespread interest in promoting traditional health systems to satisfy primary health care needs. This is often extremely true in South Asia, as prices of recent medicines spiral and governments find it increasingly difficult to make happy the worth of pharmaceutical-based health care. Medicinal plants form the proposal of traditional or indigenous systems of health in employment by the bulk of the population of most developing countries. In China, India and other countries in South and East Asia, the traditional systems of drugs are used with thousands of plant species to treat malaria, stomach ulcers, and various other disorders. Throughout the region, there's strong and sustained public support for the protection and promotion of the cultural and spiritual values of traditional medicines.

Use of plants as medicines

The use of plants as medicine is older than recorded history as mute witness to the present fact marsh mallow root, hyacinth, and yarrow is found carefully tucked round the bones of a Stone Age of civilization in Iraq. In 2735 B.C., the Chinese emperor Shen Nong wrote an authoritative treatise on herbs that's still in use today. Shen Nong, a philosopher of the era recommended the utilization of Ma Huang (known as ephedrine within the western world), for instance, which acted against respiratory distress wherein ephedrine, extracted from ephedrine, is widely used as a decongestant. In its synthetic form, there are wide use of pseudoephedrine, in many allergy, sinus, and cold-relief medications produced by large pharmaceutical companies.

Distribution of medicinal plants throughout India and World was given in Table 1.

Table 1: Distribution of medicinal plants

Country/Region	Total number of native species in flora	No of medicinal plant species reported	% of medicinal plants
World	297000	52885	10
India	17000	7500	44
Indian Himalayas	8000	1748	22

Herbal medicines

World Health Organization (WHO) estimated that 4 billion people, 80% of the planet population, presently use herbal medicine for a few aspect of primary health care. Herbal medicine might be a major component altogether indigenous peoples' traditional medicine and a standard element in Ayurvedic, homeopathic, naturopathic, traditional oriental and Native American Indian medicines. World Health Organization noted that of 119 plant-derived pharmaceutical medicines, about 74% are utilized in modern medicine in ways in which correlated directly with their traditional uses as plant medicines by native cultures. Status of various herbal based medical systems in India was given in Table 2.

Table 2: Status of various herbal based medical systems in India

Characteristics	Medical Systems				
	Ayurveda	Siddha	Unani	Tibetan	Homeopathy
Medicinal plants known	2000	1121	751	337	482
Licensed pharmacies	8533	384	462	-	613
Hospitals	753	276	74	-	223
Dispensaries	15193	444	1193	-	5634
Registered practitioners	438721	17560	43578	-	217460
Under graduate college	219	6	37	-	178
Post graduate college	57	3	8	-	31

Sources of medicinal plants

Major pharmaceutical companies are currently conducting extensive research on plant materials gathered from the rain forests and other places for his or her potential medicinal value. Altogether the countries of South Asia, medicinal and aromatic plants (MAPs) play a big role within the subsistence economy of the people, especially those living within the rural interiors. The gathering, simple processing and trading of medicinal plants contribute significantly to the cash income of the poor and ladies in these regions. A recent study administered by CECI-India, indicated that from one district of pithoragarh in Uttarakhand state of India, quite 1300 plenty of MAPs are collected and traded annually, most of them though are not legally. Unsustainable and enormous scale harvesting of MAPs from the natural habitats without providing equitable benefit to the local people and government is of grave concern to all or any. Therefore, by sustainably using and growing economically remunerative MAPs, there's an ample scope to take care of both the agricultural livelihoods and environmental sustainability. MAP-based local micro-enterprises also can bridge the gap between rural poor and comparatively well-off urban rich and promote social harmonization and sound environment conservation.

Constraints and opportunities

Some of the rational applications integrating medicinal plants into traditional farming systems have taken an obligated relationship in backstopping upland agriculture or farming. South Asian states have rich and diverse traditions of practicing complex and rotational farming systems that has herbal plants cultivation and thus, conservation and ex-situ cultivation of medicinal plants especially applying organic farming protocols features a great scope especially to access international markets. Other important opportunity and advantage of cultivating MAPs include the simplicity of their incorporation within the existing cropping systems which could be thanked to availability of an outsized number of species and selection of plant types *i.e.,* trees, shrubs, forbs, vines and their suitability to grown in several eco-physical conditions. Hence, wild stock management and cultivation of carefully selected species as a mixed, inter or companion crop in agro and farm forestry conditions is possible and wishes to be pursued. However, so as to make sure an honest input and repair delivery system including marketing cultivation may have to be administered in selected pockets in an intensive manner.

Spreads of medicinal plants in Eastern Region

The eastern region comprising of eastern UP, Bihar, Jharkhand, West Bengal, Assam, Orissa, and Chhattisgarh occupies about 22.5 per cent of the country's geographic area and is inhabited by about 35 per cent of the country's population. The region are often divided into 3 distinct physiographical units, namely (i) Plains of eastern UP, Bihar, West Bengal, and Assam; (ii) Hilly and plateau regions in eastern UP, Bihar, Jharkhand, West Bengal, Orissa, Chattisgarh, and Assam; and (iii) Coastal plains of West Bengal and Orissa.

This region features a sizable amount of valuable medicinal plants naturally growing mostly in fragile ecosystems that are predominantly inhabited by rural poor and indigenous communities. The sustainable management of those traditionally used plants not only help conserve nationally and globally important biodiversity but also provide critical resources to sustain livelihoods. In North Bihar plains and forests, for instance, these were discovered in abundance, a various range of herbs, shrubs, trees and vines that have significant medicinal value whose healing properties are known to the local healers and traditional doctors for hundreds of years but currently these are threatened due to lack of concerted conservation efforts, *e.g.* value of medicinal raw materials. Though annually exported materials from Nepal to India and other countries is estimated higher. The Eastern region could also enjoy the recent interest of the western world on growing medicinal plants in these areas bordering Nepal and earn tons quite a lot taking advantage of the liberal trade regime which has opened newer markets for these products. The eastern region of the country, recognized as a 'low productivity high potential' region needs holistic management of land, water, crops, biomass, horticultural, livestock, fishery and human resources. The region, though endowed with rich natural resources has lacked in capitalizing on these resources. Though, it's an upscale resource base for intensive and diversified agriculture, still, declining per capita land and water availability, reduced diversion of water for agriculture to satisfy the food requirement of burgeoning population including degrading land and water resources are a significant challenge towards improving productivity and sustainability. There is an urgency to get and to identify suitable crops and develop high yielding, disease and pest resistant, location-specific sorts of important, cereals, oilseed, pulses, fruits, vegetables and medicinal and aromatic plants to satisfy the worldwide competition. Growing of medicinal plants have got the following several pertinent perspectives.

Social perspective

Use of medicinal plants in primary health care and nutrition needs is traditional and imbedded altogether cultures. No major problems of acceptability regarding familiarity with the usage of plant products, methods of cultivation of the many commonly grown plants and technologies required for processing into items of common household uses and value. Med-plants have also been wont to develop family-based health and livelihood oriented enterprises in rural areas. Medicinal and Aromatic Plants (MAPs) usually helped in: (a) Preserving the normal medical knowledge and (b) Provide easily adaptable enterprising opportunities for unemployed youth and rural poor who can learn the trade from their parents and peers and earn not only their livelihood but also contribute to the society. The urgency and wish to guard the fast disappearing medicinal plants-based tradition, which remains abundant in Eastern Region, can't be overemphasized. If proper values are often added to the normal medical knowledge-based health practices and subsistence-oriented MAP applications, an outsized number of jobs are often created within the rural areas. Even at current level of conversion of traditional medicinal knowledge into economic opportunities, enterprise-based application can account for thousands of jobs in rural areas could be created. Thus, med-plants have high potential in creating jobs and pushing economic process upward in resource-constrained areas affected by limited educational opportunities, lack of infrastructure, and underdeveloped med- plants-based trade and commercial activities. The conversion of socio-cultural traditions and indigenous knowledge into livelihood means and economic opportunities also has the advantage of preserving the rapidly eroding cultural knowledge and practices which are increasingly threatened due to globalization and homogenization of individuals and communities.

Environmental perspective

The growing apathy toward products made up of chemical products becoming ethically unacceptable. This has created new markets for quality, certified and organic herbal products. Medicinal plants have the potential to fill these needs as they supply green health alternatives and variety of other eco-friendly products of domestic and industrial usage. These are found as trees, shrubs, grasses and vines, these plant species abundantly growing within diverse agro-climates and entered into the planet food and drug market because of nature of environment eco-friendly status as botanical products which looked upon as an emerging and new opportunity. The event of medicinal plants-based economic incentives is being increasingly applied to enlist greater participation of individuals in conservation of forest ecosystem.

Potential of equitable commercialization

The medicinal plants sub-sector has immense potential because the sustainable commercialization which can benefit farmers and industry both by providing higher price and by opening up national and global markets for brand spanking new products from the region of origination. Private sectors stand to profit by ensuring sustainable supply of quality raw materials to profit their industry and trade if they will be facilitated to create partnerships with farmers. Many of the species are shade tolerant are climbers, trees, shrubs and herbs which will be grown in several configurations of crop geometry.

National trade and enterprise development

The demand for medicinal and aromatic plants has come forward to satisfy both domestic and export market comprising 162 species; those are predicted to extend at about 15 to 16% between 2002 and 2005 in India. Medicinal and aromatic plants cultivation and management, therefore, can become highly remunerative both in financial and economic terms for the small-scale growers. The current gap between demand and supply is estimated to be 40,000 to 200,000 tons, which is predicted to rise from 152,000 to 400,000 tons by 2010 (Planning Commission, 2000). Not only the plants are in increasing demand by major herbal drug industries as an important staple of their drugs, but also its collection, production, processing, packaging, and transportation requires high labor input, which may create employment opportunity within the job-starved country like India. Collection from wild and selective harvesting additionally to primary processing is usually done manually, and even at the secondary and tertiary levels, med-plants have substantial labor requirements. Moreover, not only do MAP-based industries expand jobs, enhancing traditional uses through value-added processing can increase cash earnings to the local people

World market trends

European marketplace for herbal supplements is estimated very huge amount and increased over the years for herbal remedies. Germany is far and away the most important market. The market is growing rapidly at over 4% once a year for herbal remedies and considerably faster for herbal supplements. The US herbal market is nearing saturation and is predicted to peak point reached in the next few years. Demand for medicinal plants is predicted to still expand rapidly, fuelled by the expansion of sales of herbal supplements and remedies. Their basic uses in medicine will continue within the future, as a source of therapeutic agents, and because the staple base for the extraction of semi-synthetic chemical compounds like cosmetics, perfumes, and food industries.

Domestication and cultivation

Some of the rational applications integrating medicinal and aromatic plants into traditional farming systems have taken an obligated relationship in backstopping upland agriculture. Cultivation of medicinal plants are carefully selected species as a mixed, inter or companion crop in agro and farm forestry conditions following a soil-improving crop rotation is very feasible livelihood enhancing activities. However, this may require an improved input and repair delivery system including marketing, and post harvest technologies. Cultivation must be done on a business platform by a sequence of small and micro-enterprise-based groups and individuals. So as to realize an economy of scale and desired impact, it's going to got to be concentrated in selected pockets in an intensive manner as a cluster of activities and micro-enterprises. Information on the propagation of medicinal plants is out there for fewer than 10% and agro-technology is out there just for 1% of the entire known plants globally. This trend shows that developing agro-technology should be one among the thrust areas for research. Furthermore, so as to satisfy the escalating demand of medicinal plants, farming of those plant species is imperative. Apart from meeting demand, farming may conserve the wild genetic diversity of medicinal plants wherein farming permits the assembly of uniform material, from which standardized products are often consistently obtained.

Cultivation also permits better species identification, improved internal control, and increased prospects for genetic improvements. The choice of planting material for large-scale farming is additionally a crucial task. The planting material therefore should be of excellent quality, rich in active ingredients, pest- and disease-resistant, and environmental tolerant. Large scale farming, one has got to determine whether monoculture is that the right due to cultivate all medicinal plants or one have got to promote a poly-culture model for better production of medicinal plants.

Studies conducted on the agro-forestry of medicinal plants elsewhere suggest that since many medicinal plant species like better to grow under forest cover, agro forestry offers a convenient strategy for his or her cultivation also as conservation through:

1. Integrating shade-tolerant medicinal plants as lower strata species within the multi-strata system;
2. Cultivating short cycle medicinal plants as intercrops in existing stands of tree crops; and
3. Growing medicinal trees as shade providers and boundary markers, and
4. Inter-planting medicinal plants with food crops.

Hence, it's understood that the cultivation of medicinal plants isn't a simple task because the history of medicinal plants farming as reflected. Many farmers within the trans-Himalayan region of northern India have replaced the medicinal plants farming with common crops like peas (*Pisum sativum*), and potatoes (*Solanum tuberosum*) due to the lengthy cultivation cycle of medicinal plants. The value of the many medicinal plants in northern India is less than many seasonal vegetables, which may be an explanation for scanty farming of medicinal plants. However, attempts be made by different organizations to cultivate various medicinal plant species, including rare and endangered categories.

Agro-technology

Agro-technology for about 20 species of rare and endangered medicinal plants of northern India has been developed by different organizations. However, the per hectare cost of cultivation, total annual production and cost-benefit ratio found to be fluctuated with different medicinal plant species. At the present, however, the farming of most of the medicinal plant species is being operated on a really small scale and is restricted to a couple of hectares of land in various states of eastern India. There's also a market uncertainty, fear about lower prices, and lack of data about the required pre and post-harvest processes, and wishes for obtaining permits from government agencies for the cultivation of medicinal plants. Additionally, many farmers are unaware of the agencies liable for issuing permits of growing such plants as crop. If the farmers aren't granted permits needed to cultivate, they're forced to sell their products on the illegal market, which exposes them to action by government agencies and therefore the exploitation by middlemen.

Opportunities in developing the medicinal plant's sector

'Herbal industry possesses an upscale diversity of medicinal plant species across the varied forest types along an altitudinal gradient within the use and variety of medicinal plants. Such a high diversity of medicinal plants would be helpful for further research project on exploring their medical efficacy, value addition, and use in curing various old and new diseases. India has already established a reputation as a low-cost manufacturer of high-quality generic drugs within the global market. This fact is often used as a crucial tool for the marketing of herbal products produced in India. It's expected that India's aim to create a golden triangle between traditional medicine, modern medicine, and modern science are going to be a boon for developing traditional herbal medicine and therefore the medicinal plant's sector.

The way forward

Forest Departments of the eastern states need to identify two major areas; namely the conservation area and the developmental area. The conservation areas are going to be selected supported their rich medicinal plant diversity and marked for in-situ conservation and complete protection within the concerned. Within the developmental areas, aside from the protection of the prevailing bio-resources, the medicinal plant species of the neighboring areas also be introduced and cultivated at an outsized scale. The remaining areas would remain open for sustainable harvesting of the medicinal plant species. The various policies at the national and state level and their subsequent implementation would provide a chance within the advancement of the medicinal plant's sector. This model of conservation and cultivation of the medicinal plants could also be useful for generating the staple for the 'Herbal Industries' also as for ensuring the conservation of the rare medicinal plants.

Institutional support

In India, many government and non-government organizations have had focused attention on improving the medicinal plant's sector. Opportunities for funding are created to help the one that is willing to figure and to create the capacity of the medicinal plant's sector. Consistent with the mandate of the projects those could be submitted for funding within two major schemes: *viz.,* a promotional scheme, and a commercial scheme.

The main thrust areas within the promotional scheme with objectives as follows:

1. Survey and inventory of medicinal plants;
2. In-situ conservation and ex-situ cultivation of selected medicinal plants;
3. Production of quality planting material;
4. Diffusion of data through education and communication;
5. Promotion of worldwide and domestic market system, and
6. Strengthening research, development and manpower.

The commercial scheme, the main thrust areas are as follows:

1. Bulk production of medicinal plants and ensuring the availability of quality planting material;
2. Expansion of selected medicinal plants farming areas;
3. Value addition in harvesting, processing, and marketing of medicinal plants, and
4. Developing innovative marketing mechanisms/

Stakeholder's participation

Local communities, especially weaker and marginalized groups or ethnic minorities got to be involved in planning, designing, development, and implementation of the research activities and learning studies within the project. The NGOs and GOs should consult and work with the community-based organizations and have interaction them within the participatory process to involve collectors, producers, and traders including ultimate users, women, and disadvantaged groups in project implementation. In each of the chosen villages or communities, stakeholders represented CBO, NGO or PRI should implement and/or supervise the execution of projects. Their involvement from the very beginning of the project development process is predicted to reinforce people's participation within the project and supply benefits to a good range of users.

Marketing perspective

In order to know the complex market and marketing related issues, market-related information, scooping of problems & opportunities, successful case studies with a `good practice' tag got to be surveyed and studied to develop a marketable product portfolio. The tool proposed is that the value-chain or Production-to-Consumption & Marketing (PC&M) model. The result of those studies is often useful to plan equitable commercialization, identify potential small & micro-enterprises, and assess available staple resources and merchandise mix.

Existing policies

The efforts by different agencies, to promote the medicinal plants could be summarized in the following manner:

1. Assessment of Local Conditions and Requirements
2. Assessing the Demand for Medicinal and Aromatic crops
3. Training and Capacity Building
4. Role of Supporting Individuals and Institutions
5. Development of Relevant Literature in Local language
6. Emphasis on Group Approach
7. Organizing the Growers
8. Cultivation through the Organized Sector and linking the unorganized farmers
9. Pricing Mechanism and Buy Back Arrangement with partner industry;

Aromatic plants suitable for cultivation in Eastern Region

1. Lemon grass (*Cymbopogon fleuosus*)
2. Palma rosa (*Cymbopogon martinii)*
3. Java citronella (*Cymbopogon winterianus*)

4. Mentha (*Mentha arvensis*)
5. Tulsi (*Ocimum basilicum*)
6. Vetiver (*Vetiveria zizanioides*)
7. Geranium (*Pelargonium* graveolens)
8. Patchouli (*Pogostemon patchouli*)

Table 3: A few medicinal plants found locally in Eastern India and their industrial products

Name of plants	Industrial Products
Sarpagandha (*Rauvlofia serpentina*)	Ajmaline
Nayantora (*Catharanthus roseus*), Sarpagandha (*Rauwolfia serpentine*)	Ajmalicine
Bishalyakarani (*Atropa belladonna*); Dhatura (*Datura stramonium)*	Atropine
Chahgoch (*Camellia sinensis*)	Caffeine
Javacitronala or Ganbirina (*Cymbopogon winterianus*)	Citronellal
Javacitronala or Ganbirina (*Cymbopogon winterianus*)	Citronellol
Agnisikha or Uluchandan (*Gloriosa superba*)	Colchicine
Haldhi (*Curcuma longa*)	Cucurmin
Javacitronala (*Cymbopogon winterianus*)	Dimethyloctane
Dalchini, Tejapat etc. (*Cinnamomum spp.)*	Eugenol
Geranium (*Pelargonium graveolens*)	Geraniol
Bishyalakarani (*Barleria lupulina)*	Hyoscyamine
Bandar kekoa (*Mucuna prurita*)	L-Dopa
Podina (*Mentha citrata*)	Linalool
Podina (*Mentha arvensis*)	Linalyl acetate
Pani tengeshi (Marselia minuta)	Marsilin
Podina leaves (*Mentha spp.)*	Menthol
Omita (*Carica papaya*)	Papain
Sarpagandha (*Rauvlofia serpentina*)	Rescinnamine
Sarpagandha (*Rauvlofia serpentina*)	Reserpine
Mekuri Kendu (*Strychnos nux-vomica*)	Strychnine
Chahgoch (*Camellia sinensis*)	Theobromine
Chahgoch (*Camellia sinensis*)	Theophylline
Birina (*Vetiveria zizanioides*)	Vetiverol
Birina (*Vetiveria zizanioides)*	Vetiveryl acetate
Sadabahar or Periwinkle (*Chtharanthus roseus*)	Vinblastin
Sadabahar or Periwinkle (*Chtharanthus roseus)*	Vincristine

Source:http://assamagribusiness.nic.in/agriclinics/Entrepreneurship%20dev.%20through%20 medicinal.pdf

Link with bio-partnership and rural communities with the industry

Links established with growers, traders, processors, and consumers at different levels during a value- chain or production-to-consumption system framework. Equitable Bio-partnership arrangements between processing and marketing; health-care companies and community-based organizations developed to make sure dependable markets for the producers and quality supply for the industry. Industries have shown interest in direct collaboration with producer groups/associations and lots of are committed to fair and ethical commercialization. Cultivation might not be economical if a medicinal plant species are abundant within the wild and simply collected. Therefore, the less abundant species within the wild should be promoted for large-scale cultivation. Farming of any medicinal plant species should be brought into practice only after the reliable cultivation technology of the concerned species is out there. An outsized variation in climatic and soil conditions in eastern India sustain a spread of medicinal plant species, which can be cultivated consistent with their niche. Currently, there are varieties of herbs that are utilized in curing diseases but aren't documented intimately due to scarcity of communication and comparatively low frequency of their uses. In order to develop the fully-grown medicinal plants sector, there's an urgent need for documentation of indigenous uses of medicinal plants with following objectives:

1. Develop and improve the agro-technology for valuable medicinal plants, Officially recognize and protect the customary laws of indigenous people, Prepare a transparent policy for granting permits for cultivation within the stipulated time;
2. Conduct regular research and training on better harvesting and processing techniques;
3. Investigate various pathological agents infecting medicinal plants, Setup community-based management of medicinal plants farming and marketing,with analyze the market policies and
4. Certify staple for internal control;
5. Prepare a transparent policy for granting permits for cultivation within the stipulated time;
6. Conduct regular research and training on better harvesting and processing techniques,
7. Monitor and evaluate the status of medicinal plants with the help of local communities;
8. Conserve the critical habitats of rare medicinal plant species, and share benefits judiciously arising from local people's knowledge of medicinal plants.

Table 4: Medicinal Plants Commercially Suitable for cultivation in Eastern Region

Botanical name	Common name	Parts used	Nature of demand	Gestation period	Cultivation location	Demand as in 2004-05 (MT)
Andrographis paniculata	Kalmegh	Whole	Both, Dom/Exp	Short	Region wide	2197.3
Asparagus racemosus	Shatawari	Roots	Both, Dom/Exp	Short	Region wide	16658.5
Bacopa monnieri	Brahmi	Entire plant and leaves	Both, Dom/Exp	Short	Region wide	6621.8
Boehaavia diffusa	Punarnava	Stem	Both,Dom/Exp	Short	Region wide	3373.2
Centella asiatica	Brahmi	Whole	Both,Dom/Exp	Short	Region wide	134.5
Chlorophytum arundinaceum/ boerivilianum	Safed musli	Bulbs/ tuberous roots	Both, Dom/Exp	Short	Region ide	NA
Gloriosa superba	Shankhpushpi	Roots	Export	Short	Region wide	100.5
Glycyrrhiza glabra	Liquorice or Jethi madhu	Roots	Both, Dom/Exp	Short	Region wide	1359.8
Gymnema sylvestre	Madhunashini	Leaves, roots	Both,Dom/Exp	Short	Region wide	80.70t
Piper longum	Pippali	Fruits	Both,Dom/Exp	Short	Region wide	6280.4
Phyllanthus amarus/ P. niruri	Bhoomi amla	Roots	Both, Dom/Exp	Short	Region wide	2985.3
Rauwolfia serpentina	Sarpagandha	roots	Export	Short	Region wide	588.7
Tinospora cordifolia	Giloe	Roots, stem and leaves	Export	Short	Region wide	2932.6
Withania somnifera	Ashwagandha	Dried roots	Both, Dom/Exp	Short	Region wide	9127.5

Source:https://idl-bnc.idrc.ca/dspace/bitstream/123456789/27286 /9/g-120 936. pdf

Conclusion

Proper documentations are required to reduce dependency on wild resource base and generate alternative income opportunities for the agricultural and underprivileged communities. The medicinal plant's sectors are often improved if the agricultural support agencies would come to assist strengthens the medicinal plant growers and if research institutions would help the plant growers by improving their basic knowledge about cultivation practices. Awareness and interest of farmers, supportive government policies, assured markets, profitable price levels, access to simple and appropriate agro-techniques, and availability of trained manpower are a number of the key factors for successful medicinal plant cultivation.

The diffusion of any available knowledge domain on medicinal plants should be made operational by a network structure of communication. The normal uses of low profile and lesser-known medicinal plants should even be documented to disseminate their therapeutic efficacy by preparing well acceptable medicines and also to scale back the pressure on over-exploited species.

The knowledge of herbal medicines for the complicated treatment of diseases is confined to mostly the practicing herbalists or plant scientists with the assumption that herbal medicines will lose their potency if revealed to people. Although some herbs may have medicinal values, sometimes the medicinal preparations inflict certain side effects. The special significance of medicinal plants within the treatment of diseases relates to the phytochemical present. The study showed that each one the medicinal plants contain bioactive components referred to as phytochemicals. Therefore, the therapeutic or medicinal applications of those plants might be thanks to the presence of bioactive components. This communication problem limits a researcher's capability to affect the farmers' problems. Hence, communication links between researchers, extension services of institutions, and farmers should be strengthened. On many occasions, the gathering of planting material, especially of rare and endangered medicinal plant species from natural habitats for various experimental purposes by researchers, also poses a threat to their natural populations in wild. The researchers must remember of the germination potential, seedlings, and rhizomes' survival strategies of the specified species collected from the wild for scientific experiments. Researchers must plant an identical number of people back in nature after completion of research work on the collected species. There should not be any communication problem between researchers and farmers.

Some pertinent short questions

1. Name five common medicinal crops grown in plain;
2. Why Should the Eastern Region go the Herbal Way?
3. What do you know about domestication of medicinal crops?
4. Role of stakeholder in growing medicinal crops;
5. Write down the importance of institutional support for growing medicinal crops;

13

Gender Equity in Ancient and Modern Agriculture

Introduction

Agriculture is that the back-bone of the economics of eveloping country likes India. It's additionally the most "culture" for rural families even nowadays agriculture and allied activities build the one largest contribution to gross domestic product. Female are mother of the race and liaison between generations. Female's role in agriculture and agricultural heritage described and recognized by the United Nations Organization (UNO) that has sustained the expansion of society wrought the long run of states. Within the rising advanced social state of affairs female have a role to play. Rural females are the foremost vital productive hands within the economy of majority of the developing nations together with India. Female are concerned all told the aspects of agriculture. India incorporates a preponderantly rural economy. Seventieth of her population is rural; of these households, have interaction in agriculture as their main supply of financial gain. It's forever been India's most vital economic sector. Female plays a significant role during this vital agricultural sector, as a result of it's for the most part a ménage enterprise. Female in India are major producers of food in terms important, volume and variety of hours worked. Nearly 63% of all economically active men are engaged in agriculture as compared to seventy eight per cent of female. Nearly 50% of rural female employees are classified as agricultural labourers and thirty seventh as cultivators. Concerning 70% of farm work was performed by women. In several countries, women play a major and crucial role in agricultural development and agricultural allied fields together with, main crop production, live-stock production, gardening, post-harvesting operations, agro/social biology, fishing etc. as it is evident; it's a reality long taken without any consideration however unheeded since ages. Hence, here a try has been created (i) to review the role of female in agriculture and changes in their position at small level, that's village level (ii) to review the gender discrimination in land, labour and wage market.

A large variety of female employees are found in agriculture and their contribution to the agricultural economy is important, however they need not received adequate attention. Female during this sector encounter several difficulties thanks to their distressing and pitiable operating conditions like long operating hours, low wages, ill-treatment and gender discrimination. The efforts on concentration on the plight of the female employees in agriculture and rural development programmes, investigates the role of female, their migration issues, their standing within the home and society and review completely different welfare programmes introduced by the govt. of and NGOs to ameliorate and alter the lifestyles of female employees in Agriculture in India. India incorporates a national tradition certain to agriculture fertility. Within the North, the Indus depression and river region are crucial agricultural areas graced by the Ganges River and monsoon season. Supported years 2011 UNO envisaged, only 17.5% of India's gross domestic product (GDP) are accounted for by agricultural production. However for a majority of the country, associate calculable seventy two of the 1.1 billion folks that boarded rural India, it's the way of life. Agriculture defines family tradition, social relations and gender roles. Female within the sector, whether or not through ancient means that or industrial, for subsistence or as associate labourer, represents a big demographic cluster. Problems like economic independence, decision-making talents, Organization and access to education and health services and this way has created externalities like economic condition and social process, and combined problems with gender difference is tied with agriculture.

Extreme environmental condition changes are amongst the factors that have begun to jeopardize agricultural production globally. India's agricultural sector that depends greatly on the variations in climate and weather is outlined in the main by the monsoon season. The acceptable levels of precipitation that last from Gregorian calendar month to September, predicate a bountiful agricultural yield soon within the year 2014. Increasing temperatures and erratic precipitation has begun to exhaust agricultural land and make high variations of land. Within the past number of years these trends have created a lucid impact in India, inflicting droughts and unpredictable rain. Only one season of such weather patterns will be devastating to the resource of farmers, United Nations Organization will notice no resilience in little fares.

The loss of diversity in India and specifically food crops could be a serious concern of food security and property of the agricultural sector in India. The affiliation between female farmers and environmental health isn't merely for subsistence and survival. It additionally stems from an extended existing cultural valuation of India's agricultural fertility in ritual and observe. Women's affiliation to land is mirrored in their almanac-like information of plant varieties. Rituals and ceremonies in varied elements of the country show this shut relationship.

There is *Lohri*, the harvest pageant of geographic region or *navadhanya* puja that translate to the worship of 9 cereals, celebrations that occur in southern India. Each ceremony celebrates the role of female in agriculture and fertility and importance of setting and diversity.

The ancient agricultural weighs heavily used by female subsistence farmers boast environmentally friendly options, as it was found like seed preservation, natural fertilizers and crop rotation techniques that don't exhaust delicate soil. Within the wake of inexperienced Revolution's reforms, it's clear that a lot of the high yield recommendations had severe environmental impacts. The negative environmental impacts of the revolution are barely setting out to show their full have an effect on. The widespread chemical pollution in communities that utilize pesticides and herbicides is making a public pathological state that has disproportionately wedged female. In the state of geographic region that was touted as a hit of revolution cancer rates have skyrocketed. A study by Punjabi University ascertained that a high rate of genetic injury amongst farmers, that was attributed to chemical use. Mental object on the acceptable use of pesticides, leading to the serious use, improper disposal, the utilization of pesticides as room containers, and contamination of water with serious metals are causative On a small level female are organizing into collectives to exchange information, organize organic seed sharing, to pursue organic and property agricultural practices.

Success on small finance light-emitting diode by female cluster in Bangladesh has been discovered by the laureate, Prof. Yunus established the planet illustrious Gramin Bank in Bangladesh.

Women in agricultural heritage

Women play a significant role in the agricultural labour force and in agricultural activities, although to a varying degree. Consequently their contribution to agricultural output is undoubtedly extremely significant, although difficult to quantify with any accuracy. It has often been claimed that women produce 60-80 percent of food. However, assigning contributions to agricultural outputs by gender is problematic because in most agricultural households both men and women are involved in crop production. It can be attempted to allocate output by gender by assuming that specific crops are grown by women and others by men and then aggregating the value of women's and men's crops to determine the share grown by women. In rural India, the percentage of women who depend on agriculture for their livelihood is as high as 84%. Women make up about 33% of cultivators and about 47% percent of agricultural labourers. These statistics do not account for work in livestock, fisheries and various other ancillary forms of food production in the country. In 2009, 94% of the female agricultural labour force was engaged in crop cultivation were in cereal production, while 1.4% worked in vegetable production, and 3.72% were engaged in fruits, nuts, beverages, and spice crops. Women's participation rate in the agricultural sectors is about 47% in tea plantations, 46.84% in cotton cultivation, 45.43% growing oil seeds and 39.13% in vegetable production. These crops require labour-intensive work, the work is considered quite unskilled. Women also heavily participate in ancillary agricultural activities. According to the Food and Agriculture Organization, Indian women represented a share of 21% and 24% of all fishers and fish farmers, respectively. Despite their dominance of the labour force women in India still face extreme disadvantage in terms of pay, land rights, and representation in local farmers' organizations. Furthermore, their lack of empowerment often results in negative externalities such as lower educational attainment for their children and poor familial health. In India, the typical work of the female agricultural labourer or cultivator is limited to less skilled jobs, such as sowing, transplanting, weeding and harvesting, that often fit well within the framework of domestic life and child-rearing. In cotton seed production, they are engaged in pollination in the process of hybridization activity which requires patience and more precision. Many women also participate in agricultural work as unpaid subsistence labour.

Women authorization

Empowerment within the literature refers to the act of bestowing power and authority on some one. Thus, female authorization refers to the conferring of leverage to female United Nations Organization are otherwise underprivileged. This includes granting to female effectual decision-making power/authority and therefore the power to influence others selections together with economic, social and civil freedom. Authorization, by it's terribly definition, implies a rise within the ability to exercise power. In India, as in most developing countries, female are believed to be and treated as inferior to men. Moreover, as the lives of female and men are embedded in a very matrix of unequal gender relations, a decrease within the gender difference is important for associate outcome of 'empowerment' for girls. In different words, changes like augmented financial gain, skills and authority is also higher understood as enablers that promote women's authorization. However, the women, so as to be considerably scattered to attain their perceived goal, first of all would like authority reception, that in most cases they are doing not have. The method ought to, therefore, be administered at the same time reception and out of doors. Inside the family they need to have equal say as men so ought to be the case at work. The foremost in depth part of female authorization is providing them with social station, standing and justice. Major attributes that contribute to female authorization are education, social equity and standing, improved health, economic or money stability and political participation. In India, a banging 56% of the female are illiterate as against a significantly 24% just in case of men, evincing the putting difference. This must be considerably increased in a very 5 year timeframe. Educating the woman kid is currently associate integral a part of the proper to Education Act effective that ought to; therefore, considerably enhance the female acquisition level at par with men in a very 5 year timeframe. This apart adult acquisition programmes ought to be initiated in villages to contribute to the education in female acquisition.

Self-facilitated teams of female

Self-facilitate teams are term as SHGs were first fashioned by KVK Gadag with a pair of, 167 members from five blocks. It's additionally conducted coaching programmes for the members on conception of SHG and thrift besides composing exposure visits to sure-fire SHGs. The thrift and credit activities of the SHGs embody Rs 28.50 hundred thousand savings by the SHGs throughout the year and Internal disposition of Rs 22.00 hundred thousand to the members.

Twenty-six coaching programmes on varied financial gain generating activities were conducted by the KVK for 599 SHG members. Several of the SHGs were joined to the money establishments and different government agencies (SC / ST Development Corporation, Zilla council etc.) for money help. The entire monetary fund of the teams was Rs 66.38 lakh. The financial gain generating activities haunted by the SHGs enclosed hand-crafted product, bakery product, little business, tailoring/embroidery, goat/buffalo rearing and vermin-composting. Under Mission Mode NATP authorization of female in agriculture, 540 farm female were classified into thirty six Self facilitate teams for beginning completely different enterprises supported the requirements and preference of farm female, resources offered and selling potentiality within the space. Trainings were organized for capability building of farm female of the SHGs within the enterprises. The members of Self facilitate teams were additionally trained to handle completely different equipment's. Authorization of female Self facilitate teams are created by talent coaching and orientation to project management aspects.

Innovation of SHGs

The KVK initiated the conception of Saturday Bazaar in Gadag city to encourage the SHG started enterprises as per their interest. The female have started generating financial gain from enterprises. Under the project Involving rural female in cultivation : A step towards making certain economic and nutritional security fifty six grounds ponds of Puri and Khurda district were hand-picked for fish culture and nursery raising through the active involvement of rural female. Baseline data of the taking part female, water analysis of

The pools and trainings on pond preparation, nursery rearing of rohu and catla, improvement of weeds, removal of weed fishes, manuring and lime application were undertaken. Nursery rearing of rohu and catla spawn for one to one ½ month made healthy and quality fries.

Women within the agricultural labour force: The terms "economically active in agriculture" and "agricultural labour force" are identical ideas. This idea is broader than employment in agriculture as a result of it includes those fired however probing for add the world.

Two styles of knowledge will contribute to activity the contribution of female within the agricultural labour force: statistics on the share of female within the economically active population in agriculture and time use surveys, that document the time spent by men and female in numerous activities.

Female-economically active population in agriculture: Data on the economically active population in agriculture are offered for several countries, and supply the foremost comprehensive line of the participation of female in agriculture. During this live, a private is according as being within the agricultural labour force if he or she reports that agriculture is his or her main economic activity. However, these knowledge might underestimate female participation in agriculture for reasons mentioned below, and caution is suggested in deciphering changes over time as a result of enhancements in knowledge assortment is also liable for a number of the ascertained changes. According to this knowledge, female comprise simply over 40% of the agricultural labour force within the developing world, a figure that has up slightly since 1980 and ranges from concerning 20% within the Americas to nearly 50% in Africa. Even considering this knowledge as lower bounds for the participation of female within the agricultural labour force, they are doing not support estimates on top of 60% aside from many countries.

The global female average is dominated by Asia. Inside Asia, the sub-regional averages vary from concerning 35% in South Asia to nearly 50% in East and Southeast Asia. The Asian average is dominated by China, wherever the female share of the agricultural labour force has augmented slightly throughout the past 3 decades. The female share in India has remained steady at simply over 30% these terribly giant countries mask changes in some smaller countries wherever the female share of the agricultural labour force seems to own augmented dramatically, currently exceptional 50% in Bangladesh. Different Asian countries like Asian country have seen declining female labour shares in agriculture.

Women conjure nearly 50% of the agricultural labour force in Sub-Saharan Africa, a rise from concerning 45% in 1980. The averages in Africa vary from simply over 40% in Southern Africa to only over 50% in East Africa. These sub-regional averages have remained fairly stable since 1980, with the exception of Northern Africa, wherever the female share seems to own up from 30% to nearly 45% The sub-regional knowledge for Africa conceal wide variations between countries each within the share of female labour in agriculture and therefore the trend. The developing countries of dry land have abundant lower average female agricultural labour shares than the opposite developing country regions at simply over 20% in 2010, slightly over in 1980. The South Yankee countries of Bolivia, Brazil, Colombia, Ecuador and Republic of Peru dominate the common and are liable for most of the rising trend.

Female summon 60-80% of the agricultural labour force: The statement is commonly seen that female represent 60 to 80% of the agricultural labour force in developing countries. This datum appears to own originated in associate early study from the international organization Economic Commission for Africa that states: "Few persons would argue against the estimate that female are liable for 60-80% of the agricultural labour equipped on the continent of Africa...." (UNECA, 1972). A decade later, variety of country statements in a very report from the Food and Agriculture Organization according that female represent between 70% and 90% of the agricultural labour force in several desert African countries (FAO, 1984). Similar statements are still being created nowadays for all developing countries (Action Aid; Gates Foundation reality sheet). though the offered knowledge show that female play a major – though extremely variable - role within the agricultural labour force, the estimates according on top of don't represent current conditions within the cluster of developing countries as a full. Several researchers have questioned the gender patterns that emerge from the utilization knowledge given here and on top of (see for instance Beneria, 1981). Industrialist (2005) identifies variety of potential sources of idea of female employment in labour markets, and in agriculture especially. She notes that rural female in Latin America geographical area seemingly to reply that "their home" is their primary responsibility, although they're heavily engaged in agriculture. Different difficulties arise as a result of censuses tends to stress income-generating activities - so underestimating subsistence production - and since agricultural production is commonly outlined as munitions. Activities like rearing little ethereal mammal, room agriculture, and post-harvest process are usually undercounted. Industrialist focuses on critiquing the numbers for geographical area, however similar criticisms are valid for different regions, like South Asia.

Time spent in agricultural activities: Time use surveys plan to offer a lot of complete account of your time use by men and female than are offered from the labour force statistics according on top of. Such studies typically aren't nationwide representative and aren't directly comparable as a result of they sometimes cowl little samples, report on differing kinds of activities (that aren't forever clearly specified) and use completely different methodologies. Despite these caveats, an outline of the proof from studies that specify time use by agricultural activity suggests fascinating patterns.

Time-use surveys of all agricultural activities, revealed appreciable variation across countries, and typically inside countries, however the info are generally almost like the labour force statistics mentioned on top of. In Africa, estimates of the time contribution of female to agricultural activities ranges from concerning 30% within the African nation to 60-80% in numerous elements of Cameroon. In Asia, estimates vary from 32% in India to over 50% in China. This is lower in geographical area; however exceeds 30% in some elements of Republic of Peru. Two separate studies are according every for African nation and Republic of Peru, and variations replicate completely different time periods and locations inside the countries.

A putting degree of within-country variation is shown by time-use knowledge for India. whereas the nationwide representative knowledge indicates that the national average for women's share of total time-use in agriculture is 32% knowledge for state and Rajasthan reports women's share as from but 10% to over 40%, severally. However in each areas, younger female women contribute a better share of the entire time provided in agriculture by their cohort than older women kill theirs. In Rajasthan, for instance, women between fourteen -15; Distinction noted between Rajasthan and state may well be as a result of a distinction within the total variety of hours provided by female or as a result of a distinction within the total variety of hours provided by men. 9 to 90 ineteen years older contribute up to 60% of the entire time spent on agriculture by their group. Time-use studies additionally revealed that female time-use in agriculture varies wide betting on the crop and therefore the part of the assembly cycle, the age and grouping in question, the sort of activity and different factors. Knowledge from state reveals larger involvement of female in upland rice than wet rice and within the management of young plantation crops like cinnamon and rubber instead of constant crops at maturity.

They offer policy-relevant data concerning wherever, once and the way to focus on interventions geared toward female and the way to bring men into the method constructively. Given the variability in gender roles in agriculture, generalizations concerning time use from one region to a different aren't applicable. Studies that contemplate the gender roles inside their specific geographic and cultural context will offer sensible steerage for policy manufacturers and practitioners concerned in technology investments, extension services, post-harvest activities and selling interventions. A few of the time use studies gift the precise breakdown of your time use by farming activity. 5 of the studies, covering six cases, data on 5 common categories: land preparation, chemical application, weeding, gather and storage. All studies with the exception of Bangladesh (where, for rice, women's time is sort of entirely spent on post-harvest activities) found weeding to be a preponderantly female activity, followed by gather (in three studies) and chemical application (2 studies). Female were generally concerned all told different activities (except for ploughing) however didn't offer a majority of labour. 6 from Bangladesh, Ghana, African nation, Vietnam, India and Asian nation additionally for the case of tomato contract farming within the country gather could be a preponderantly female activity.

Finally, there was a tendency to note that the time-use studies that collected the relevant knowledge ensure the popular perception that female overpoweringly offer the best proportion of ménage time spent on food process and preparation. If these aspects of food preparation are enclosed, women's labour share might well exceed 60% in several African countries and will approach 60% in several Asian ones.

Women's contribution to agricultural production

The flower industry in Latin America provides an interesting case in contrasting points of view. In Colombia, for example, finds that 64 percent of the workforce directly growing fresh-cut flowers for export is women and consider this type of agro-industrial work skilled, while others consider it unskilled. Women do have supervisory jobs among those directly involved in cultivation activities, they have a much lower share of managerial or professional jobs in other aspects of the sector similarly, it was found that in sectors producing primarily for the export market, and women tend to be replaced by males as profits increase.

The arrival of the flower industry in the Ecuadorian town of Cayambe in the late 1980s (in combination with other household and individual factors) affected time use patterns in some surprising ways. The total time spent by women in paid and unpaid work did not increase, contrary to a frequent criticism of agricultural export development which maintains that women are unduly burdened by work in the industry. Indeed, the most compelling evidence of the industry's impact was on men's increased participation in housework. In Cotocachi, Ecuador, in contrast, women were not prepared to move or even commute to work in the flower industry despite the higher wages offered there. The women did not view flower employment as an option, indicating either that their husbands would not allow them to work or that the work would be detrimental to family relations.

In Senegal, the growth of modern horticulture supply chains has been associated with direct beneficial effects for rural women and reduced gender inequalities in rural areas. That study also finds that women benefit more from employment in large-scale estate production and agro-industrial processing than from high-value smallholder contract-farming in which they often provide unpaid family labour;

Conclusion

The empirical evidence were found that on women's roles in agriculture, setting the stage for subsequent analysis on gender differences in agriculture and the potential gains from removing these gender differences. The main findings are:

Women comprise about 43 percent of the global agricultural labour force and of that in developing countries, but this figure masks considerable variation across regions and within countries according age and social class. Women comprise half or more of the agricultural labour force in many African and Asian countries, but the share is much less in some.

Time use surveys, which provide a more comprehensive assessment of how men and women spend their time, further emphasize the heterogeneity among countries and within countries in women's contribution to agriculture. The labour burden of rural women exceeds that of men, and includes a higher proportion of unpaid household responsibilities related to preparing food and collecting fuel and water. The contribution of women to agricultural and food production is clearly significant. However, it is impossible to verify empirically the share produced by women because agriculture is usually a venture among household members and involves a range of resources and inputs that cannot be readily assigned by gender. Women's participation in rural labour markets show much heterogeneity at the regional level, but women are over represented in unpaid, seasonal and part-time work, and the available evidence suggests that women are often paid less than men, for the same work.

Hence, the exact regionally specific information and analysis is necessary for good gender-aware agricultural policy making. Data collection has improved substantially over the last decades, as has our understanding of the complexity of women's roles and the need to collect data not only on primary activities but on all women's activities. Data are needed to better understand gender roles in agriculture and how they change over time and in response to new opportunities. Women's roles are diverse and that they vary across regions and countries. These roles cannot be understood properly, and interventions targeting cannot be designed effectively, without also understanding their differential access to land, capital, assets, human capital, and other productive resources.

Some pertinent short questions

1. What is gender equity in agriculture?
2. What are the roles of female in agricultural operations?
3. What is the women's role in self help group formation?
4. What is SHGs and how it is formed?
5. Write down the importance of women as agricultural labour force;

14

Crop Significance and Crop Classifications

Introduction

The crops, its significance along with its classification has got paramount importance in agricultural heritage. A crop is a plant or animal product that can be grown and harvested in an area with a target to earn profit or achieve subsistence. Crop may refer either to the harvested parts or to the harvest in a more refined state (husked, shelled, etc.). Most crops are cultivated in agriculture or aquaculture.

A crop is usually expanded to include macroscopic fungus (e.g. mushrooms), or alga (alga-culture) at the point of origination. Most crops are harvested as food for humans or livestock (fodder crops) or to get raw materials for agro-based industry. Some crops are gathered from the wild collection (including intensive gathering). The important non-food crops include horticulture, floriculture and industrial crops. Horticulture crops include plants used for other crops (e.g. fruit trees, grown single or in orchard). Floriculture crops include blanket plants, houseplants, flowering garden and pot plants, cut cultivated greens, and cut flowers. Industrial crops are produced for clothing (fiber crops), bio-fuel (energy crops, algae fuel), or medicine (medicinal plants).

Microbes (fungi, bacteria or viruses) are rarely referred to as crops. Animals raised for human or animal consumption are referred to as livestock and microbes as microbiological cultures. Microbes are not typically grown for food itself, but are rather used to alter food (e.g., for production of citric acid, fermenting yogurt, soya sauce and so on). The importance of a crop varies greatly by region. Crop production depends on the availability of arable land and is affected in particular by yields, macroeconomic uncertainty, as well as consumption patterns; it also has a great incidence on agricultural commodities' prices. The importance of crop production is related to harvested areas, returns per hectare (yields) and quantities produced. Crop yields stands for harvested production per unit of harvested area for the particular crop i.e.

products. In most of the cases yield data are not recorded, but are obtained by dividing the production data by the data on area harvested. The actual yield that is captured on farm depends on several factors such as the crop's genetic potential, the amount of sunlight, water and nutrients absorbed by the crop, the presence of weeds and pests infestation. These indicators are presented for wheat, maize, rice and soybean. Crop production is measured in tons per hectare, in thousand hectares and thousand tons.

Today, India ranks second worldwide in farm output. India is an agrarian country and about 60% of population depends on agriculture for their livelihood. The people residing in urban areas may not realize much importance of agriculture but this fact is not new that agriculture is the main source of income for major part of our country's population. India's population is growing faster than its ability to produce rice and wheat. The required level of investment for the development of marketing, storage and cold storage infrastructure is estimated to be huge. The country produces innumerable crops ranging from medicinal to cereal. These commodities are used for various purposes from human consumption, in industries, for animal feed and so on.

In spite of the impressive achievements, the Indian agricultural sector continues to face poor infrastructure conditions. About 40 per cent of the cultivated land is under assured irrigation system. Farmers on the remaining 60 per cent of the land are completely dependent on vagaries of annual rainfall, which is also greatly characterized by large variations in terms of precipitation both spatially and in time. A large majority of farmers in different parts of the country gains from application of science and technology in agriculture have yet to be realized. As a result, the productivity levels of many major crops do not compare very favourably with the yields obtained in agriculturally advanced countries. Further, these factors coupled with high illiteracy constrain the farmer's ability to shift to more remunerative cropping patterns in response to market signals. Therefore, their capacity to take advantage of the opportunities presented by liberalization of trade is limited. The country's agriculture has gained in strength and resilience since independence, although growth in agriculture is highly skewed over regions and crops. However, the agriculture sector in India is now faced with intense internal and external pressures arising from the impact of policies of economic liberalization. Efficient and effective management of agriculture will be crucial in the years to come for acquiring enduring self-reliance and ensuring sustainable growth with an emphasis on consideration of equity.

Farming continues to be the major source of food, nutrition, income and employment for the most of the rural population. In India, farming is characterized by presence of a large number of small and marginal scale farmers with small land holdings. However, the country is blessed with diverse agro-climatic conditions which enable the farmers to produce a large number of agricultural commodities. The challenge of producing enough food for the growing population with the reducing holdings is a great task. With the development of commercial agriculture techniques during the post-independence period the agriculture sector has been able to cater to the domestic and international markets. In the light of the focus on commercial farming, the rich tradition of crop diversity of Indian farming remains with its glory. Earlier crop diversity was more prominent; however, after Green Revolution in 1960s country witnessed erosion of diversity. At present, few crops and few varieties and hybrids of those crops are occupying major production area and are grown repeatedly year after year in succession. This has resulted in emergence of several field levels biotic and abiotic constraints, yield plateauing and overall reduction in the benefits realized from farming. To overcome these problems as well as to reach the goal of production sustainability, crop diversification may be an option as because diversity assures soundness of ecology.

Causes of crop diversification

Crop diversification provides the farmers with a wider choice in the production of a variety of crops in a given area so as to expand production related activities on various crops and also to minimize the possible risk. Crop diversification in India is generally viewed as a shift from traditionally grown less remunerative crops to more remunerative crops. In recent years, the crop diversification is also taking place due to governmental policies, thrust on some crops, market reforms, infrastructure development, government subsidies, price support mechanisms, higher profitability and stability in production. Crop diversification and growing of large number of crops are practiced in dry-land areas to reduce the risk factor of crop failures due to climatic abnormality and recurring droughts. Crop substitution and crop shift are also taking place in the areas suffering with some specific soil related problems.

The country has made considerable progress in the farm sector during the last 70 years. From 'hand to mouth' conditions in the early sixties, the country has not only become self-reliant in food grains but has acquired sufficient resilience to tide over the adverse conditions. The achievements in food production is the outcome of a policy framework of improving rural infrastructure including irrigation, research, extension, provision of agricultural inputs at reasonable prices, mechanization in farming, marketing support through minimum support price mechanism and promotion of farmer producer organizations (FPOs). No doubt, it has been realized that such development has been made at a cost of degradation of crop ecosystem. Though an impressive achievement has been made in Indian agricultural sector farming continues to face poor infrastructure conditions and vagaries of climate change. For a large majority of farmers in different parts of the country gains from application of science tools and technologies in agriculture have yet to be realized and resource poor small and marginal farmers of dryland areas are still adopting subsistence farming. As a result, the productivity levels of many major crops in India do not compare very favourably with the yields obtained in agriculturally advanced countries. Due to limited capacities of the farmers to take advantage of the opportunities presented by liberalization is further limited. Efficient and effective management of agriculture is a crucial aspect in the years to come for acquiring enduring self-reliance and ensuring sustainable growth.

The farmers in the developing countries try to grow several crops in their holdings in an agricultural year. The level of crop diversification largely depends on the geo-climatic, socioeconomic conditions and technological development in a region. In general, higher is the level of agricultural technology, lesser the degree of diversification. Moreover, it has been observed that the rich farmers go with specialized agricultural enterprise while the poor and subsisted farmers of fragile ecological regions are generally showing more interest in crop diversification.

Opportunities in advanced agro-technologies

It is being increasingly realized that agriculture is no longer a subsistence activity carried out by peasant cultivators, but rather an enterprise and manufacturer of biomass using land, water, genetic material and the latest in technology. The agriculture of the twenty first century will increasingly be farmers' entrepreneurship harnessing technologies to optimize returns from his land and investments he makes on it. Biotechnology and genetic engineering in crops with focus on primary productivity and also on many quality traits will go a long way to improving the yield and quality of many important crop plants. With the advent of such emerging technologies and consequent scope for increased economic returns, the diversification in favour of such crops will be the future focus. Many other related technologies and their adoption will also inject an added dimension in crop diversification. Decision support systems, governmental policies, geographic information system, application of information technology leading to market information etc., will also lead to crop diversification primarily on economic considerations.

Strategies for crop diversification

Realizing the importance of crop diversification, the central and different state governments have taken several initiatives. Horticulture sector has been given highest importance considering its importance in nutrition security. Some of the important programs of the government include:

(a) Launching of National Horticulture Mission;

(b) Launching of Technology Mission for the Integrated Development of Horticulture in the North-Eastern Region;

(c) Implementing Agriculture Insurance Scheme (namely PMFBY);

(d) Operationalzing Technology Mission on Cotton;

(e) Provision of Capital Subsidy for construction/modernization/expansion of cold storages and storages for horticultural produce;

(f) Creation of Watershed Development Fund at the National level for the development of rain-fed areas;

(g) Infrastructure Support for Horticultural Development;

(h) Strengthening Agricultural Marketing;

(i) Seed and Gene Bank Schemes; and

(k) Cooperative Sector Reforms.

Measurement of crop diversification, there is developed a formula based on the gross cropped area. The formula has been expressed as:

Index of Crop Diversification = Percent of sown area under x crops /Number of x crops

Where x crops are those crops that individually occupy 10 per cent or traditionally of the gross cropped area in the area under study.

The main advantage of crop diversification is that it provides a relationship between the relative a real strength of the crops grown in region. The larger the number of crops having about 10 % of gross cropped area, the higher is the crop diversification. In fact, it is an indicator of multiplication of agricultural activities which obviously involve intense competition amongst various activities for space. The keener the competition, the higher the degree of diversification, and lesser the competition, greater will be the degree of specialization, or monoculture. The degree of crop diversification is closely influenced by the soil characteristics, soil moisture, amount of rainfall received, the availability of irrigation facilities, the accessibility of the arable land and the technology deployed by the cultivators. As stated earlier, the areas of extreme wet or extreme dry climate are least conducive for crop diversification. India, being a vast country of continental dimensions, presents wide variations in agro-climatic conditions. Such variations have led to the evolution of regional niches for various crops. Historically, regions were often associated with the crops in which they specialize for various agronomic, climatic, hydro-geological, and even, historical reasons. But, in the consequences of technological changes encompassing bio-chemical and irrigation technologies, the agronomic niches are undergoing significant changes. With the initiation of irrigation and new farm technologies, the yield level of most crops-especially that of cereals has witnessed an upward shift making it possible to obtain a given level of output with reduced area or more output with a given level of area and creating thereby the condition for inter-crop area shift (diversification) without much disturbance in output level. Besides, as agriculture become drought proof and growth become more regionally balanced, there has been a reduction in the instability of agricultural output. A comprehensive plan for each of the region of high diversification may go a long way in enhancing their agricultural productivity and in reducing the regional inequalities in agricultural development. There is increasing awareness amongst the experts of agriculture that crop diversification with suitable crop rotations is necessary for the maintenance of soil health and for making agriculture more productive and sustainable.

Crop classifications

It is well known that there are more than 600 cultivated plant species, from which there are about 200 species play important role in the world trade. However, about only fifteen field crops as well as plant species represent the most important economic crops. Therefore, these crop species must be classified or grouped in a convenient way to facilitate communication, dissemination and retrieval of scientific information as well as promotes the conservation, and improvement of certain plants. Generally, classification of these species is important for these reasons:

1. To get acquainted with crops;
2. To understand the requirement of soil & water different crops;
3. To know adaptability of crops;
4. To know the growing habit of crops;
5. To understand climatic requirement of different crops;
6. To know the economic produce of the crop plant and its use;
7. To know the growing season of the crop; and
8. To know the actual condition required to the cultivation of plant; and

In general, field crop is an organism grown and/or harvested for obtaining yield. Agronomic point of view, crop is a plant in single or multiple cultivated for economic purpose. Classification is done to generalize similar crop plants as a class for better understanding of them.

Classification types used in field crops

Based on ontogeny (Life cycle)

Based on economic use (Agronomic)

Based on Botany (Scientific)

Based on seasons

Based on climate

Based on special purpose

Based on life span

Based on root depth

Based on CO_2 fixation

Mode of pollination

Based on leaf morphology

Based on cultural requirements

Based on origin

Center of origin

1. Based on ontogeny (Life cycle)

• Annual crops

Crop plants that complete life cycle within a season or year, called annual. They produce seed and die within the season. Example, wheat, rice, maize, mustard and others;

• Biennial crops

Plants that have life span of two consecutive seasons or years. First years/ season, these plants have purely vegetative growth usually confined to rosette of leaves. The tap root is often fleshy and serves as a food storage organ. During the second year / season, they produce flower stocks from the crown and after producing seeds the plants die. Example, sugar beet, beet root, and others;

• Perennial crops

They live for three or more years. They may be seed bearing or non-seed bearing. Ex. Napier fodder grass, coconut, and others;

2. Based economic use (Agronomic)

Cereals: Cereal, the derived from word 'Ceres' which denotes as 'Goddess' who was believed as the giver of grains by Romans. Cereals are the cultivated grasses grown for their edible starchy grains. Larger grains used as staple food like rice, wheat, maize, barley, and oats and so on. Cereal grain contains 60 to 70% of starch and is excellent energy rich foods for humans. Some important cereals are:

1. Rice: *Oryza sativa*
2. Wheat: *Triticum aestivum*
3. Maize: *Zea mays*

Millets: Millets are small and coarse grained cereals, staple food in drier regions of the developing countries are called millets. They are also annual grasses of the group cereals. But' they are grown in lesser area or less important area whose productivity and economics are also less important. These are also staple food for people of poor countries. In India, pearl millet is a staple food in Rajasthan.

Millets are further broadly classified in to two, as major millets and minor millets.

Major millets

1. Sorghum *(Jowar): Sorghum bicolor*
2. Pearl millet *(Bajra): Pennisetum glaucum*
3. Finger millet *(Ragi): Eleusine coracona*

Minor millets

1. Foxtail millet: *Setaria italica*
2. Little millet: *Panicum miliare*
3. Common millet: *Panicum miliaceum*
4. Barnyard millet: *Echinochloa colona* var *frumentaceae*
5. Kodo millet: *Paspalum scrobiculatum*

Pulse: Seeds of leguminous plants used as food grain rich in protein. Pod containing grain is the economic portion. Pulses are preferred for protein rich value & also economically important in cropping system. The wastes or stalk is called the 'haulm' or 'stover'. Haulm is used as green manure and has high value cattle feed. Green pods used as vegetables, e.g. cowpea, lablab. Seed coat of pulses are nutritious cattle feed. Some of the examples of pulse crops are as follows.

1. Red gram: *Cajanus cajan*
2. Black gram: *Vigna mungo*
3. Green gram: *Vigna radiata*
4. Cowpea: *Vigna unguiculata*
5. Bengalgram: *Cicer arietinum*
6. Horsegram: *Macrotyloma uniflorum*
7. Lentil: *Lens esculentus*
8. Soybean: *Glycine max*
9. Pea: *Pisum sativum*
10. Grass pea - *Lathyrus sativus*

Oilseeds: Crops which are rich in fatty acid are cultivated for the production of vegetable oil. They are used either for edible or industrial or medicinal purposes.

1. Groundnut or peanut: *Arachis hypogeae*
2. Sesame or gingelly: *Sesamum indicum*
3. Sunflower: *Helianthus annuus*
4. Castor: *Ricinus communis*
5. Linseed or flax: *Linum usitatissimum*
6. Niger: *Guizotia abyssinia*
7. Safflower: *Carthamus tinctorius*
8. Indian Mustard: *Brassica juncea*

Sugar crops**:** Crops cultivated for sugar. Juice is extracted from stem of sugarcane used for jaggery or sugar. Number of by products like molasses, bagasse, pressmud etc. is obtained from sugar industry. Molasses used for alcohol and yeast formation and bagasse for paper making and fuel. Pressmud used for soil amendment; whereas, trash (green leaf and dry foliage) is used for cattle feed. Sugar beet is another sugar crop where tubers are mainly used for extraction of sugar. Tubers and tops are used as a fodder for cattle feed.

1. Sugarcane: *Saccharum officinarum*
2. Sugar beet: *Beta vulgaris*

Fibre crops Plants are grown for obtaining fibre.

Different kinds of fibre are,

i) seed fibre, cotton; ii) Stem/ bast fibre, Jute, mesta; iii) leaf fibre, *Agave sp.*

1. Cotton: *Gossypium hirsutum*
2. Jute: *Corchorus olitorius/ C. capsularis*
3. Mesta: *Hibiscus cannabinus*

Forage: Forage refers to vegetative matter, fresh or preserved, utilized as feed for animals.

1. Grasses: Bajra (*Pennisetum glaucum),* napier *(Pennisetum purpureum*), guinea *grass (Megathyrsus maximus),* para grass *(Brachiaria mutica)*
2. Legumes: Lucerne (*Medicago sativa*), Barseem *(Trifoloim alexandrium)*

Spices and condiments: Crop plants or their products used for flavour, taste and add colour to the fresh or preserved food.

1. Ginger: *Zingiber officinale*
2. Garlic: *Allium sativum*
3. Fenugreek: *Trigonella foenum-graecum*
4. Cumin: *Cuminum cyminum*
5. Turmeric: *Curcuma longa*
6. Coriander: *Coriandrum sativum*
7. Anise: *Pimpinella anisum*

Medicinal plants: Crops used for preparation of medicines. Example,

1. Tobacco: *Nicotina tabacum*
2. Peppermint: *Mentha piperita*
3. Aswagandha: *Withania somnifera*
4. *Aloe: Aloe vera*

Beverages: Products of crops used for preparation of mild, agreeable and simulating drinking.

1. Tea: *Camellia sinensis*
2. Coffee: *Coffea arabica*
3. Cocoa: *Theobroma cacao*

3. Based on scientific or botanical classification

Botanical classification is the family and scientific names of plants which consist of genus and species and are universally accepted. Carolos Linnaeus, a Swedish botanist, was responsible for the binomial system of classification. Botanical classification is based upon similarity of plant parts and flower structure. This is the most important way of classification because it determines to what extent the plants are relatives. Field crops belong to the "**spermatophyte**", or seed plant, division of "**plant kingdom**", which includes plants reproduced by seeds. Within this division, the common crop plants belong to the subdivision of "**Angiosperm**", which are characterized by producing seeds with coats (covered seed). The "**angiosperm**", are then divided into two classes, namely, **monocotyledons** and the **dicotyledons**. All the grasses, which include the cereals and sugarcane, are **monocotyledons**. The legumes and other plants except the grasses are classified as **dicotyledons**. Each of these two classes is still further divided into orders, **familes**, **genera**, **species** and **varieties**.

4. Based on Binomial system

Each crop plant has two names; the genus and the species. The genus starts with capital letter, whereas the species name is small. The name of the man who first gave the accepted name is indicated by a letter or abbreviation. For example, the scientific name of wheat is *Triticum aestivum* L. The letter L. signifies that the name of Swedish botanist Carl Linnaeus who named it. Sometimes a crop can be renamed by another scientist. Therefore, the crop name is followed by the initial of the first person (between bracts) followed by the initial of the second scientist as in soybean. The scientific name of this crop is *Glycine max* (L), Merr.

According to the botanical classification we can summarize the families of the most important field crops as follows:

- Monocotyledons

Poaceae/ Gramineae: wheat, barley, rice, maize, oat, sugar cane, sorghum, rye grass, sudan grass and different millets.

Liliaceae: onion and garlic.

- Dicotyledons

Leguminosae: field bean, lupine, check pea, lentil, fenugreek, Egyptian clover, alfalfa, soybean, peanut, grass pea, caster bean, red clover and white clover.

Malvaceae: cotton.

Linaceae: flax.

Solanceae: potato, tobacco.

Pedaliaceae: sesame.

Composite: sunflower, safflower.

5. Based on seasons

Crops are grouped under the seasons in which their major field duration falls.

Kharif crops: Crops grown during June-July to September-October which require a warm wet weather during their major period of growth and shorter day length for flowering. Ex. Rice, maize, castor, groundnut.

Rabi crops: Crops grown during October–November to January-February, which require cold dry weather for their major growth period and longer day length for flowering. Ex. Wheat, mustard, barley, oats, potato, Bengal gram, berseem, cabbage and cauliflower.

Summer crops: Crops grown during February–March to May–June which require warm dry weather for growth and longer day length for flowering. Example, black gram, greengram, sesame, cowpea;

6. Based on climatic condition

Tropical crop: Coconut, sugarcane

Sub-tropical crop: Rice, cotton

Temperate crop: Wheat, barley

Polar crop: All pines, pasture grasses

7. Based on special purpose

These classifications are used to refer to plants having special advantages to the farmer himself in relation to his farming practices, and include:

Catch or emergency crops: These crops are used to substitute crops that have failed on account of unfavorable conditions. They are usually quick-growth crops, such as rye, millet and clover. In Egypt, clover can be grown and one cut can be obtained before planting cotton crop.

Cash crop: any short maturing crop which is grown to generate income while the main crop is still in its vegetative stage of growth; any crop grown to generate cash rather than for subsistence. Some crops may be cash crops one year but not the next, or for one farmer but not another.

Cover crops: these crops are planted to provide a cover for the soil and to prevent erosion such as clover and rye.

Green manure crops: these crops are turned under while still green in order to improve the soil properties and increase organic matter content. Several field crops can be used such as Egyptian clover, lupine and cowpea.

Companion crops: in this case a crop can be intercropped with another one and each crop is harvested separately. For example, onion and garlic can be intercropped with cotton crop, or soybean with maize.

Silage crops: these crops are preserved in a succulent condition by partial fermentation in a tight receptacle. They include corn, sorghum, forage grasses and legumes.

8. Based on life span

All field crops can be divided into three categories according to the length of their life cycle as follows:

Annual crops: plants of this category complete their entire life cycle from seed to seed in a single growing season and then die. Most field crops are considered annual crops such as wheat, barely, rice, maize, sorghum, faba-bean, lentil, check pea, lupine, flax, soybean, sesame, sunflower, safflower, and others.

Biennial crops: these plants complete their life cycle in two seasons. Vegetative growth occurs during the first season resulting in a rosette form but plants don't start flowering (blooming). In the second season, the green plants give flowers and seeds. The crops of this category are onion, sweet clover, and sugar beet.

Perennial crops: these crops are grown in the soil for more than two years (they can persist for more than two years). They may either produce seed or not every year. In other words, they have an indefinite life period. They do not die after reproduction but continue to grow indefinitely from year to year. Sugar cane, white clover, and alfalfa are examples of perennial crops.

9. Based on root depth

It is clear that the root system of field crops differ in structure, function and extent. Therefore, field crops can be classified according to the depth of their roots as follows:

Hallow root crops: the root system of these crops extends in the soil to a depth of one meter such as wheat, barley and rye.

Intermediate crops: the depth of the root system of these crops ranges from 1- 1.5 meter in the case of *faba* bean and sugar beet.

Deep root crops: the root system of these plants extends in the soil to a depth more than 1.5 meter as in *alfalafa.*

10. Based on CO_2 fixation

C_3 Plants: Net assimilation rate in these plants is low but compensation point is high. Stomata are open during the day. Photo respiration is high in these plants C_3 Plants have lower water use efficiency. The initial product of C assimilation is the three 'C' compounds. The enzyme involved in the primary carboxylation is ribulose-1,-Biophospate carboxylose. Example, rice, soybeans, wheat, barley cottons, potato; C_3 plants include more than 95 percent of the plant species on earth.

C_4 plants: Net assimilation rate in these plants is high but compensation point is low. Stomata are open during the day. The primary product of C fixation is four carbon compounds which may be malice acid or acerbic acid. The enzymes responsible for carboxylation are phosphoenol Pyruvic acid carboxylose which has high affinity for CO_2 and capable of assimilation CO_2event at lower concentration, photorespiration is negligible. These are said to be drought resistant & they are able to grow better even under moisture stress. C_4 plants translate photosynthates rapidly. Example, sorghum, maize, sesame, millets;

CAM plants: (CAM stands for Crassulacean Acid Metabolism) the stomata open at night and large amount of CO_2 is fixed as a malice acid which is stored in vacuoles. During day stomata are closed. There is no possibility of CO_2 entry. CO_2 which is stored as malice acid is broken down & released as CO_2. In these plants there is negligible transpiration. C_4 & cam plant have high water use efficiency. These are highly drought resistant. Example of CAM plants are pineapple, sisal and agave.

11. Based on mode of pollination

1. *Naturally self-pollinated crops:* - the predominant mode of pollination in these plants is self-pollination in which both pollen and embryo sac are produced in the same floral structure or in different flowers but within the same plant. Examples: rice, most pulses, okra, tobacco, tomato.

2. *Naturally cross-pollinated crops:* pollen transfer in these plants is from the anther of one flower to the stigma of another flower in a separate plant, although self pollination may reach 5 percent or more. Examples: corn and many grasses, avocado, grape, mango, many plants with unisexual or imperfect flowers.

3. *Both self- and cross-pollinated crops:* these plants are largely self-pollinated but varying amounts of cross-pollination occur. Examples: cotton and sorghum.

12. Based on leaf morphology

Broad leaved: Mustard, pea, potato, tobacco.

Narrow leaved: Rice, wheat, sugarcane, onion, ginger, finger millet.

13. Based on cultural requirements

A. According to suitability of tropo-sequence

Crops grown on upland: Redgram, maize, jower, pearl millet.

Crops grown on medium land: Jute, potato, sugarcane, wheat.

Crops grown on low land: Rice, para grass.

B. According to suitability of the textural group of soils

Crops of sandy to sandy loam soil (light textured): Potato, sweet potato, sugarbeet, onion.

Crops of silty to silty loam soil (medium textured): Jute, sugarcane, maize, cotton.

Crops of clay to clay loam soil (heavy textured): Rice, wheat, lentil, para grass guinea grass.

C. According to the tolerance to the problematic soil

Crops tolerant to acidic soil: Potato, mustard, wet rice.

Crops tolerant to saline soil: Chillies, sugar beet, beet, sunflower, barseem.

Crops tolerant to alkaline soil: Barley, cotton, pea, barseem.

Crops tolerant to water logged soil: Rice, para grass, napier, job's tier

Crops tolerant to soil erosion: Vetiver, moth bean, marvel grass.

D. According to tolerance to hazardous weather conditions

Frost tolerant crops: Sugar beet, beet.

Cold tolerant crops: Potato, cabbage, mustard.

Drought tolerant crops: Millets, barley, safflower, castor.

E. According to water supply

Irrigated crop: Boro rice, potato, tobacco.

Rainfed crop: Millets, upland rice, jute, castor, Dinanath grass.

Rainfed, but partially irrigated: Chick pea, mustard, maize.

Residual soil moisture crops: Tori, lathyrus, linseed.

Rainfed flooded crops: Deep water rice, makhana, water chestnut.

F. According to method of sowing or planting

Direct seeded: Upland rice, mustard, chick pea, lentil.

Planted: Potato, napier, citronella.

Tranplanted: Transplanted rice, tobacco, onion.

G. According to requirement of inter-tillage and earthing-up

Intertilled crops: Potato, sweet potato, sugar beet.

Non-intertilled crops: Lentil, oats, chick pea, linseed.

H. According to crop duration

Very short (upto 75 days): Greengram, spinach.

Short (76-100 days): Upland rice, early potato.

Medium (101-125 days): wheat, jower, bajra, barley.

Long (126-150 days): Tobacco, barseem.

Very long (>150 days): Sugarcane, red gram.

I. According to method of harvest

Reaping: Rice, wheat, mustard.

Uprooting by pulling: Linseed, radish, beet.

Uprooting by digging: Potato, sweet potato.

Picking: Cottin, maize, okra.

Cutting: Barseem, napier.

Grazing: Para grass, pasture grasses.

J. According to post-harvest operation required

Curing: Tobacco.

Stripping: Jute, mesta, sunhemp.

Shelling: Groundnut, maize.

Ginning: Cotton;

14. Based on centre of origin of crops

The origin of crop plants is now basic to plant breeding in order to locate wild relatives, related species, and new genes (especially dominant genes, sources of disease resistance). Knowledge of the origins of crop plants is vitally important in order to avoid genetic erosion, the loss of germplasm due to the loss of ecotypes and landraces, loss of habitat (such as rainforests), and increased urbanization. Germplasm preservation is accomplished through gene banks (largely seed collections but now frozen stem sections) and preservation of natural habitats (especially in centers of origin). So far eight Vavilovian centres of are there which further consist old and new world as origin of different crops.

Ancient World

I. Chinese Center: The largest independent center which includes the mountainous regions of central and western China, and adjacent lowlands. A total of 136 endemic plants are listed, among which are a few known to us as important crops.

Cereals and Legumes

Broomcorn millet (*Panicum miliaceum)*; Italian millet (*Panicum italicum*); Japanese barnyard millet (*Panicum frumentaceum*); kaoliang (*Andropogon sorghum*); buckwheat (*Fagopyrum esculentum*); hull-less barley (*Hordeum hexastichum*); soybean (*Glycine max*); adzuki bean (*Phaseolus angularis*); velvet bean (*Stizolobium hassjoo*).

Roots, Tubers, and Vegetables

Chinese yam (*Dioscorea batatas*); radish (*Raphanus sativus*); Chinese cabbage (*Brassica chinensis, B. pekinensis*); onion (*Allium chinense, A. fistulosum, A. pekinense*); cucumber (*Cucumis sativus*)

Fruits and Nuts

Pear (*Pyrus serotina, P. ussuriensis*); Chinese apple (*Malus asiatica*); peach (*Prunus persica*); apricot (*Prunus armeniaca*); cherry (*Prunus pseudocerasus*); walnut (*Juglans sinensis*); litchi (*Litchi chinensis*).

Sugar, Drug, and Fiber Plants

Sugarcane (*Saccharum sinense*); opium poppy (*Papaver somniferum*); ginseng (*Panax ginseng*); camphor (*Cinnamomum camphora*); hemp (*Cannabis sativa*).

II. Indian Center: This area has two sub-centers.

A. Main Center (Hindustan): Includes Assam and Burma, but not northwest India, Punjab, nor northwest Frontier Provinces. In this area, 117 plants were considered to be endemic.

Cereals and legumes

Rice (*Oryza sativa*); chickpea or gram (*Cicer arietinum*); pigeon pea (*Cajanus indicus*); urd bean (*Phaseolus mungo*); mung bean (*Phaseolus aureus*); rice bean (*Phaseolus calcaratus*); cowpea (*Vigna sinensis*).

Vegetables and tubers

Eggplant (*Solanum melongena*); cucumber (*Cucumis sativus*); radish (*Raphanus caudatus*); taro (*Colocasia antiquorum*); yam (*Dioscorea alata*).

Fruits

Mango (*Mangifera indicia*); orange (*Citrus sinensis*); tangerine (*Citrus nobilis*); citron (*Citrus medica*); tamarind (*Tamarindus indica*).

Sugar, oil, and fiber Plants

Sugar cane (*Saccharum offi cinarum*); coconut palm (*Cocos nucifera*); sesame (*Sesamum indicum*); safflower (*Carthamus tinctorius*); tree cotton (*Gossypium arboretum*); oriental cotton (*Gossypium nanking*); jute (*Corchorus capsularis*); crotalaria (*Crotalaria juncea*); kenaf (*Hibiscus cannabinus*).

Spices, Stimulants, Dyes, and Miscellaneous

Hemp (*Cannabis indica*); black pepper (*Piper nigrum*); gum arabic (*Acacia Arabica*); sandalwood (*Santalum album*); indigo (*Indigofera tinctoria*); cinnamon tree (*Cinnamomum zeylanticum*); croton (*Croton tiglium*); bamboo (*Bambusa tulda*).

III. Indo-Malayan Center: Includes Indo-China and the Malay Archipelago. Fifty-five plants were listed, including:

Cereals and Legumes

Job's tears (*Coix lacryma*); velvet bean (*Mucuna utilis*)

Fruits

Pummelo (*Citrus grandis*); banana (*Musa cavendishii, M. paradisiaca, H. sapientum*); breadfruit (*Artocarpus communis*); mangosteen (*Garcinia mangostana*).

Oil, sugar, spice, and fiber plants

Candlenut (*Aleurites moluccana*); coconut palm (*Cocos nucifera*); sugarcane (*Saccharum offi cinarum*); clove (*Caryophyllus aromaticus*); nutmeg (*Myristaca fragrans*); black pepper (*Piper nigrum*); manila hemp or abaca (*Musa textilis*).

IV. Central Asiatic Center: Includes Northwest India (Punjab, Northwest Frontier Provinces and Kashmir), Afghanistan, Tadjikistan, Uzbekistan, and western Tian-Shan. Forty-three plants are listed for this center, including many wheats.

Grains and Legumes

Common wheat (*Triticum vulgare*); club wheat (*Triticum compactum*); shot wheat (*Triticum sphaerocoecum*); pea (*Pisum sativum*); lentil (*Lens esculenta*); horse bean (*Vicia faba*); chickpea (*Cicer arientinum*); mung bean (*Phaseolus aureus*); mustard (*Brassica juncea*); flax (*Linum usitatissimum*), sesame (*Sesamum indicum*).

Fiber Plants

Hemp (*Cannabis indica*); cotton (*Gossypium herbaceum*).

Vegetables

Onion (*Allium cepa*); garlic (*Allium sativum*); spinach (*Spinacia oleracea*); carrot (*Daucus carota*).

Fruits

Pistacia (*Pistacia vera*); pear (*Pyrus communisi);* almond (*Amygdalus communis*); grape (*Vitis vinifera*); apple (*Malus pumila*).

V. Near-Eastern Center: Includes interior of Asia Minor, all of Transcaucasia, Iran, and the highlands of Turkmenistan. Eighty-three species including nine species of wheat were located in this region.

Grains and Legumes

Einkorn wheat (*Triticum monococcum* 14 chromosomes); durum wheat, (*Triticum durum,* 28 chromosomes); poulard wheat (*Triticum turgidum,* 28 chromosomes); common wheat (*Triticum vulgare,* 42 chromosomes), oriental wheat (*Triticum orientale*); Persian wheat (*Triticum persicum, Triticum timopheevi,* 28 chromosomes); Wheat (*Triticum macha,* 42 chromosomes; *Triticum vavilovianum*, 42 chromosomes); two-row barley (*Hordeum distichum, H. nutans*); rye (*Secale cereal*); mediterranean oats (*Avena byzantine*); common oats (*Avena sativa*), lentil (*Lens esculenta*); lupine (*Lupinus pilosus, L. albus*).

Forage Plants

Alfalfa (*Medicago sativa*); Persian clover (*Trifolium resupinatum*); fenugreek (T*rigonella foenum graecum*); vetch (*Vicia sativa*); hairy vetch (*Vicia villosa*).

Fruits

Fig (*Ficus carica*); pomegranate (*Punica granatum*); apple (*Malus pumilo*): pear (*Pyrus communis*); Quince (*Cydonia oblonga*); cherry (*Prunus cerasus*); hawthorn (*Crataegus azarolus*).

VI. Mediterranean Center: Includes the borders of the Mediterranean Sea. Eighty-four plants are listed for this region including olive and many cultivated vegetables and forages.

Cereals and Legumes

Durum wheat (*Triticum durum expansum*); emmer wheat (*Triticum dicoccum*); Polish wheat (*Triticum polonicum*); spelt (*Triticum spelta*); mediterranean oats (*Avena byzantine*); sand oats (*Avena brevis*); Canarygrass (*Phalaris canariensis*); grass pea (*Lathyrus sativus*); pea (*Pisum sativum*); lupine (*Lupinus albusi*).

Forage Plants

Egyptian clover (*Trifolium alexandrinum*); white clover (*Trifolium repens*); crimson clover (*Trifolium incarnatum*); serradella (*Ornithopus sativus*).

Oil and Fiber Plants

Flax (*Linum usitatissimum,* L. *angustifolium*); Rape (*Brassica napus*); black mustard (*Brassica nigra*); Olive (*Olea europaea*).

Vegetables

Garden beet (*Beta vulgaris*); cabbage (*Brassica oleracea*); turnip (*Brassica campestris*), lettuce (*Lactuca sativa*); Asparagus (*Asparagus offi cinalis*), celery (*Apium graveolens*); chicory (*Cichorium intybus*); Parsnip (*Pastinaca sativa*); rhubarb (*Rheum offi cinale*).

Ethereal Oil and Spice Plants

Caraway (*Carum carvi*); anise (*Pimpinella anisum*); thyme (*Thymus vulgaris*); peppermint (*Mentha piperita*); sage (*Salvia offi cinalis*); hop (*Humulus lupulus*).

VII. Abyssinian Center: Includes Abyssinia, Eritrea, and part of Somaliland. In this center were listed 38 species.

Grains and Legumes

Abyssinian hard wheat (*Triticum durum abyssinicum*); poulard wheat (*Triticum turgidum abyssinicum*); emmer wheat (*Triticum dicoccum abyssinicum*); Polish wheat (*Triticum polonicum abyssinicum*); Barley (*Hordeum sativum*); grain sorghum (*Andropogon sorghum*); pearl millet (*Pennisetum spicatum*); African millet (*Eleusine coracana*); cowpea (*Vigna sinensis*); flax (*Linum usitatissimum*).

Miscellaneous

Sesame (*Sesamum indicum*); castor bean (*Ricinus communis*); garden cress (*Lepidium sativum*); coffee (*Coffea Arabica*); okra (*Hibiscus esculentus*); myrrh (*Commiphora abyssinicia*); indigo (*Indigofera argentei*).

New World

VIII. South Mexican and Central American Central : Includes southern sections of Mexico, Guatemala,
Honduras and Costa Rica.
Grains and Legumes

Maize (*Zea mays*); common bean (*Phaseolus vulgaris*); Lima bean (*Phaseolus lunatus*); Tepary bean (*Phaseolus acutifolius*); jack bean (*Canavalia ensiformis*); grain amaranth (*Amaranthus paniculatus leucocarpus*).

Melon Plants

Malabar gourd (*Cucurbita fi cifolia*); winter pumpkin (*Cucurbita moshata*); chayote (*Sechium edule*).

Fiber Plants

Upland cotton (*Gossypium hirsutum*); bourbon cotton (*Gossypium purpurascens*); chayote (*Sechium edule*).

Miscellaneous

Sweet potato (*Ipomea batatas*); arrowroot (*Maranta arundinaceai*); pepper (*Capsicum annuum, C. frutescens*); papaya (*Carica papaya*); guava (*Psidium guayjava*); cashew (*Anacardium occidentale*); wild black cherry (*Prunus serotina*); cochenial (*Nopalea coccinellifera*); cherry tomato (*Lycopersicum cerasiforme*); cacao (*Theobroma cacao*); tobacco (*Nicotiana rustica).*

IX. South American Center: (62 plants listed) Three subcenters are found.

A. Peruvian, Ecuadorean, Bolivian Center: Comprised mainly of the high mountainous areas, formerly the center of the Megalithic or Pre-Inca civilization. Endemic plants of the Puna and Sierra high elevation districts included:

Root Tubers

Andean potato (*Solanum andigenum,*96 chromosomes) and other endemic cultivated potato species., fourteen or more species with chromosome numbers varying from 24 to 60; edible nasturtium (*Tropaeolum tuberosum)*, edible canna (*Canna edulis*); potato (*Solanum phureja,* 24 chromosomes).

Grains and Legumes

Starchy maize (*Zea mays amylacea*); lima bean (*Phaseolus lunatus*, secondary center); common bean (*Phaseolus vulgaris,* secondary center).

Vegetable Crops

Pepino (*Solanum muricatum*); tomato (*Lycopersicum esculentum*); ground cherry (*Physalis peruviana*); Pumpkin (*Cucurbita maxima*); pepper (*Capsicum frutescens*).

Fiber Plants

Egyptian cotton (*Gossypium barbadense*).

Fruit and Miscellaneous

Passion fl ower (*Passifl ora ligularis*); guava (*Psidium guajavai*); heilborn (*Carica candamarcensis*); Quinine tree (*Cinchona calisaya*); tobacco (*Nicotiana tabacum*).

B. Chiloe Center (Island near the coast of southern Chile)

1. Common potato (*Solanum tubersum,* 48 chromosomes); wild strawberry (*Fragaria chiloensis*)

C. Brazilian-Paraguayan Center

Manioc (*Manihot utilissima*); peanut (*Arachis hypogaea*); rubber tree (*Hevea brasiliensis*); pineapple (*Ananas comosa*); Brazil nut (*Bertholletia excels*); cashew (*Anacardium occidentale*); Purple granadilla (*Passifl ora edulis*).

Conclusion

Crop significance and classification is based on crop diversification is one of the best options to increase farm income leading to food, nutrition and ecological security as well as poverty alleviation in the region. Therefore, greater attention should be paid to crop classification and crop significance those developed in a region over a considerable period of time. Crop diversification has got it two magnitude that one could be approached either by complementary and or by interactive ways on questioning the people. The crop diversification is scientifically based on crop significance and crop classification wherein diversification has either been horizontal diversification through expanding the crop base by substituting or adding more crops into the cropping systems as commonly practiced by many countries of the region; and or through vertical diversification in which downstream activities are undertaken to add value, indicating the stage of industrialization of the crops and their economic returns. Vertical crop diversification is stood with complementary to horizontal crop diversification, based on opportunities and value addition to achieve highest economic returns.

Some pertinent short questions

1. What is crop significance in agricultural heritage?
2. What s crop classification in agricultural heritage?
3. What down crop distribution and crop diversification;
4. What do you know about Indian centre of crop classification?
5. What do you know about Indo Malayan centre of crop classification?

15

National Agricultural Setup in India

Introduction

India has one of the best agricultural research systems in the world with better amount man-power amongst the developing countries. The research system includes more than 30,000 scientists along with 1,00,000 subordinate personnel actively engaged in research system in the country involved to agricultural research which considered to be second highest next to China. The achievement and success looks very much impressive and at par with all the developed nations of the world.

National Agricultural System

India has one of the largest agricultural research systems in the world with the largest number of scientific personnel of any developing country except China. The research system includes approximately 30,000 scientists and more than 100,000 supporting staff actively engaged in research related to agriculture. Although the total number of scientists engaged in agricultural research in India looks very impressive, it compares less favorably with many developed countries.

The present agricultural research system comprises essentially two main streams, the ICAR at the national level and the Agricultural Universities at the State level. Besides, several other agencies such as General Universities, Scientific Organizations, and various Ministries/Departments at the Centre, as also Private or Voluntary Organizations participate directly or indirectly in research activities related to agriculture.

1. ICAR System
2. SAU System
3. Other Agencies

1. ICAR System

The ICAR has the following major objectives: (i) to undertake, aid, promote and coordinate agricultural, animal husbandry and fisheries education, research, and its application; (ii) to act as a clearing house of research and general information relating to agricultural and veterinary matters; (iii) to maintain a research and reference library; (iv) to do other things considered necessary to attain the above objectives; and (v) to provide consultancy services in the fields of education, research and training in agriculture and allied sciences.

Amongst the major scientific organizations in the country, ICAR is unique in having concurrent responsibility for both research and education. As an apex body at the national level, ICAR is mainly responsible for the promotion and coordination of agricultural research in the various branches of agricultural and allied sciences in the country. In addition to its promoting and coordinating roles, ICAR is also directly involved in undertaking research at the national level, basic as well as applied, on diverse problems facing production of crops, animals, fisheries, etc., with the objective of evolving new production technologies suited to different agro-climatic conditions. Just as the University Grants Commission (UGC) plays a major role for the general education in the country, ICAR plays a similar role in the area of agricultural education. The charter of the ICAR also includes extension education, which is carried out through a network of projects and other mechanisms.

Research Infrastructure of the ICAR

Although agriculture is a State subject, ICAR has established many Central Research Institutions over the years to meet the agricultural research needs of the country. These are essentially meant for: (i) implementing research mandates extending beyond the administrative boundaries of the States; (ii) pursuing basic research not undertaken by most Agricultural Universities; (iii) evaluating research results through multi-location testing; and (iv) developing manpower for Agricultural Universities and other agricultural institutions.

A. Central Research Institutes: ICAR directly administers 49 research institutes in the areas of crop, animal and fishery sciences. These are:

(a) Research Management Academy: National Academy of Agricultural Research Management (NAARM) originally started as Central Staff College for Agriculture, at Hyderabad provides research management training to the agricultural scientists in the country. In addition, it organizes seminars, conferences and workshops, both national and international, based upon the scientific studies and reviews undertaken on the management problems encountered in the research system. It publishes training materials and functions as a repository of information in the field of agricultural research and education management. Besides meeting national needs as a premier management institution, the Academy has now developed an institutional capability to act as a Regional Training Centre in this part of the World

(b) National Bureau: In order to collect, conserve and initiate such measures as would lead to long-term productivity of basic resources like plants, animals, fish, soil, and water, ICAR has established four national bureau. These are:

1. National Bureau of Plant Genetic Resources (NBPGR) at New Delhi undertakes research and coordinates activities in germplasm collection; introduction and exchange of seeds and plant materials; and characterization, documentation, maintenance, and conservation of genetic resources for utilization in crop management.
2. National Bureau of Soil Survey and Land Use Planning (NBSS&LUP) at Nagpur is engaged in the preparation of soil map of India; preparation of district level soil resource inventories; soil correlation and classification at national level; research in soil genesis and classification; imparting training in soil survey and mapping; soil taxonomy, land use planning, etc; and establishment of a soil data bank for use in agricultural research and extension.
3. National Bureau of Animal Genetic Resources (NBAGR) at Karnal is engaged in the collection, maintenance and conservation of animal genetic resources for utilization in livestock improvement.
4. National Bureau of Fish Genetic Resources (NBFGR) at Lucknow is engaged in the collection, conservation and efficient utilization of fish genetic resources.

(c) Crop Science Institutes: There are 9 crop science institutes carrying out basic and applied research on specific crops and transferring the results thereof. These are:

1. Indian Agricultural Research Institute (IARI) at New Delhi is the premier agricultural institution engaged in basic and applied research in crops, postgraduate education and training, extension education, and transfer of technology. It has a Deemed University status, and awards postgraduate degrees in various disciplines of crop science.
2. Central Rice Research Institute (CRRI) at Cuttack is engaged in basic and applied research in all disciplines of rice culture; in the generation of information for planning adaptive research; and serves as a centre of information on all matters concerning rice production, protection and conservation.
3. Central Research Institute for Jute and Allied Fibers (CRIJAF) at Barrackpore is engaged in developing varieties of jute suitable for different jute growing areas; in developing appropriate crop management and crop protection technology including broad types of farm tools and

implements for jute and allied fibers; and in the production of breeder and foundation seeds of jute.

4. Central Tobacco Research Institute (CTRI) at Rajahmundry is engaged in varietals improvement of tobacco varieties; in pest and disease management; in the improvement of curing technology; in the utilization of tobacco wastes; and in extension education.
5. Indian Grassland and Fodder Research Institute (IGFRI) at Jhansi carries out basic and applied research on grasses, grass lands and fodder crops including all aspects of forage seed production and its protection for producing high quality forage. It also attempts to develop and evaluate various crop/tree species combinations for silvi-pasture and agro-forestry systems.
6. Sugarcane Breeding Institute (SBI) at Coimbatore is engaged in evaluating important sugarcane varieties for different agro-climatic regions in the country. It conducts research on breeding methodologies by taking advantage of its situational factor favourable for the flowering of sugarcane crop. It also imparts postgraduate training.
7. Central Institute of Cotton Research (CICR) at Nagpur is engaged in basic and applied research to improve cotton production; in the collection and conservation of germplasm for cotton improvement programmes; and in imparting training in advanced cotton protection technology.
8. Indian Institute of Sugarcane Research (IISR) at Lucknow has the mandate to standardize the sugarcane production and protection technologies; to devise and develop prototype of machines and implements required to promote the interest of sugarcane cultivation for maximum economic benefits to the farmers; to provide advisory services; and to impart training in sugar crops and agro-techniques.
9. Vivekananda Parvatiya Krishi Anusandhan Shala (VPKAS): Located at Almora is financed in the development of bigger top accepting and aggressive varieties of altered cereals, millets, pulses, vegetables, and fodder crops developed in hills; collection, appraisal and aliment of germplasm research of Hill crops and identification of acceptable bigger agriculture systems for these areas; and conduct research on Soil and Water Management problems and alteration research after-effects for the account of farmers of Hill areas.

(d) Horticulture and Plantation Crops Institutes: There are 6 horticulture and plantation crops institutes conducting and coordinating research on the crops they deal with. They are:

1. Indian Institute of Horticultural Research (IIHR) at Bangalore has the mandate to conduct cytogenetical studies to improve horticultural crops; standardize propagation techniques; nutritional growth regulators in horticulture; physiology and biochemistry of flower and fruit development; control of weeds, viral, fungal and bacterial diseases, mites and pests; postharvest technology; design of tools and implements; and conservation of germplasm.
2. Central Institute of Horticulture for Northern Plains (CIHNP) at Lucknow is engaged in the investigation of major production problems of fruit and vegetable cultivation for the northern plains with special reference to mango.
3. Central Institute of Temperate Horticulture (CITH) at Srinagar has the mandate to carry out basic and applied research relating to temperate fruits and vegetables in the country.
4. Central Potato Research Institute (CPRI) at Shimla has the mandate to conduct and coordinate potato research in India; to serve as a centre of information on all aspects of potato research and development; and to produce breeder seed required by the country.
5. Central Tuber Crops Research Institute (CTCRI) at Trivandrum conducts and coordinates research on all tropical tuber crops other than potato, viz. cassava, sweet potato, amorphophallus, aroids, yams, arrowroot, etc.
6. Central Plantation Crops Research Institute (CPCRI) at Kasargod has the mandate to improve the genetic potential of plantation crops; conduct basic and applied research; serve as an information centre on all matters relating to these crops; and produce genetically superior planting materials.

(e) Resource Management Institutes: There are 8 resource management institutes which are primarily responsible for undertaking research on soil and water conservation for optimizing production of crops under different conditions. They are:

1. Central Soil and Water Conservation Research and Training Institute (CSWCR&TI) at Dehradun has the mandate to study erosion problems, and conservation of land and water resources; evaluation of hydrological barriers and management of watersheds; identification of suitable plant materials for different land use; development of suitable technology for increasing production from arid lands; development of techniques for rainfed farming and efficient water management; imparting training to

state and central officers in soil and water conservation; and monitoring of changes in environment affected by integrated water management.

2. Central Soil Salinity Research Institute (CSSRI) at Karnal has the mandate to collect information on the extent, characteristic, genesis, and classification of salt affected soils; study soil and water dynamics in irrigated agriculture; conduct detailed hydrological survey; evolve methods to check deterioration of water due to pollution, and utilization of different qualities for agricultural purposes; study salt tolerance for reclamation of salt affected land; and impart postgraduate education and training.

3. Central Arid Zone Research Institute (CAZRI) at Jodhpur has the mandate to evolve location-specific technologies for optimizing production of arid lands based on ecological principles by judicious utilization of natural resources; and to train and to conduct workshops for adopting the new technologies.

4. Central Research Institute for Dryland Agriculture (CRIDA) at Hyderabad has the mandate to carry out basic research in conservation, management and utilization of natural resources in dryland ecosystem; to study the phenomenon governing crop growth and development under dryland conditions; to develop technology for exploitation of natural resources at farm level for increasing and stabilizing crop production in dryland; and also to act as a repository of knowledge on dryland farming in the country.

5. ICAR Research Complex for North-Eastern Hill Region (ICAR-NEH) at Shilling caters to the needs of agriculture, animal husbandry, fisheries, soil and water conservation, etc., for the hill areas of the north-eastern region with a major research focus on shifting cultivation, horticultural crops, pest management, livestock improvement, arid soil management, and postharvest technology.

6. ICAR Research Complex for Goa (ICAR-GOA) at Ela is engaged in research related to horticultural and other crops, livestock improvement, fisheries, etc. in the region.

7. Central Agricultural Research Institute (CARI) for Andaman and Nicobar Islands at Port Blair conducts research on high value cash and plantation crops; develops silvipastoral system and appropriate land use pattern through cropping systems; develops effective health coverage and livestock production systems; and conducts studies on capture and culture fisheries including coastal aquaculture.

8. Indian Institute of Soil Science (IISS) at Bhopal has the mandate to study the fundamental aspects of soils, particularly those that are basic to develop agricultural expertise and generate information on various basic aspects of soil research; to collaborate with other organizations to identify gaps and provide direction for further research; and to have collaboration with similar international research institutions.

(f) Technological Institutes: The technological and engineering problems in crop production and quality of commercial crops are handled by five institutes. They are:

1. Central Institute of Agricultural Engineering (CIAE) at Bhopal has the mandate for research and development of improved farm equipment related to crop production; postharvest technology; development of energy resources and power units for agriculture; to liaise with industry for the manufacture of improved implements; and to train farmers on modern agricultural technology.
2. Central Institute for Research on Cotton Technology (CIRCT) at Bombay has the mandate to improve the production of quality cotton; and to find ways and means for better utilization of cotton and its by-products.
3. National Institute of Research on Jute and Allied Fiber Technology (NIRJAFT) at Calcutta has the objectives of improvement of fiber quality; preparation of textiles by blending jute and other fibers; basic research on fibers and products; and transfer of technology for application in agriculture and industry.
4. Indian Lac Research Institute (ILRI) at Ranchi carries out research towards effecting improvements in cultivation, modification and standardization of lac, and studies its constitution and modifications so as to intensify its production and utilization; and to impart training in improved methods of lac cultivation and its industrial use.
5. Central Institute of Postharvest Engineering and Technology (CIPET) at Ludhiana undertakes research on various aspects of postharvest technology related to agricultural produce.

(g) Animal Science Institutes: Eight animal science institutes have the mandate of breeding animals for higher productivity and suggest better management practices. They are.

1. Indian Veterinary Research Institute (IVRI) at Izatnagar, a Deemed University, has its objectives to conduct basic and applied research on all aspects of livestock health, production, and nutrition; and impart postgraduate education in veterinary sciences and animal husbandry.

2. National Dairy Research Institute (NDRI) at Karnal, a Deemed University, has the mandate to meet the manpower needs for research, teaching, and dairy development through undergraduate and postgraduate instructional programmes; study of dairy production, milk and milk processing, dairy economics and management; and conduct of transfer of technology programmes.
3. Central Sheep and Wool Research Institute (CSWRI) at Avikanagar has the mandate to improve productivity of indigenous breeds of sheep through selection or cross breeding, superior exotic breeds well adapted to the tropical conditions to improve the carpet, wool and meat production.
4. Central Institute for Research on Goats (CIRG) at Makhdoom undertakes research for developing superior strains of goats for high productivity of meat, milk and fiber; to develop package of practices for feeding, management, and disease cover; and to study various aspects of goat meat, milk and pashmina.
5. Central Avian Research Institute (CARI) at Izatnagar has the objectives to develop avian species of economic importance with their optimum productivity; to conserve, evolve and improve indigenous and exotic germplasm; to impart training at various levels; to transfer the technology developed; and to undertake studies on various aspects of management to evolve environmental and ecological factors of postharvest technology.
6. Central Institute for Research on Buffaloes (CIRB) at Hissar carries out research on all aspects of buffalo production; co-ordinates research on buffalo in the country; functions as a clearing house of information on all aspects of buffalo development; establishes a nucleus breeding herd of important buffalo breeds for genetic studies on improvement of milk, meat, and draught potential; builds up adequate germplasm of improved breeds; and organizes training programmes in buffalo management.
7. National Institute of Animal Genetics (NIAG) at Karnal provides scientific support to the NBAGR and takes up research work of very fundamental nature not ordinarily taken up at the existing ICAR Institutes and Agricultural Universities.
8. National Institute of Animal Nutrition and Physiology (NIANP) at Bangalore conducts research related to nutritional and physiological aspects of livestock and management.

(h) Fisheries Institutes: Six fisheries institutes conduct studies for assessing the production of fish, conduct training programmes and undertake research. They are:

1. Central Inland Capture Fisheries Research Institute (CICFRI) at Barrackpore has the mandate to develop systems for monitoring of fish population in rivers, fresh water reservoirs, estuaries, and to study factors influencing these population and systems for optimum exploitation; and to conduct postgraduate and specialized training and extension programmes.
2. Central Marine Fisheries Research Institute (CMFRI) at Cochin conducts research for assessing and monitoring exploitable marine fishery resources for rational exploitation and conservation; to assess the exploited and under-exploited fishery resources; to understand the fluctuation in abundance of marine fishery resources; to develop suitable mariculture technology for fun fish and shell fish in open seas; and to conduct transfer of technology and postgraduate and specialized short-term training programmes.
3. Central Institute of Fisheries Education (CIFE) at Bombay has the Deemed University status. It conducts undergraduate and postgraduate degree programmes in fishery sciences; undertakes research in basic disciplines related to fish and conducts short-term and long-term training programmes for different disciplines of fishery sciences.
4. Central Institute of Fisheries Technology, (CIFT) at Cochin conducts research for the improvement of indigenous crafts and gears, and develops suitable designs for them; develops technologies for handling, processing, preservation, product development, quality control, packaging and transportation of fish and fishery products; and conducts transfer of technology and training programmes in fishery technology.
5. Central Institute of Brackish water Aquaculture (CIBA) at Madras conducts multidisciplinary, mission-oriented applied research to develop appropriate technologies for the aquaculture organisms in the estuaries, brackish water and salt intrusion areas; and to provide an information base for sustained growth and accelerated development of these fisheries through training, education and research linkages.
6. Central Institute of Freshwater Aquaculture (CIFA) at Dhauli conducts research for developing low input aqua-farming to benefit small and marginal farmers and also system of industrialized aquaculture for entrepreneurs; to improve existing technologies for carps and air-breathing fish culture, and develop culture technology for cat fish,

freshwater prawns and mussels; to increase freshwater fish production through genetic upgrading, increased pond productivity and evolution of cheap and balanced diet; to conduct nutritional and disease aspects of fish culture; and to conduct postgraduate education and training programmes.

(i) Social Science Institutes: Two institutes come under this category. They are:

1. Indian Agricultural Statistics Research Institute (IASRI) at New Delhi conducts research in experimental design, surveys, statistical genetics, computer and data processing; imparts postgraduate courses for training professional statisticians; and provides advisory and consultancy services to agricultural scientists in the country.
2. National Centre for Agricultural Economics and Policy Research (NCAP) at New Delhi has the objectives to undertake research related to the economic aspects of agricultural production process including the associated policy issues.

The ICAR institutes vary greatly in age, size, and scope. Some are old (IVRI) while some are new (IIAEPR); some are big (IARI) while some others are small (CIRG); some have a narrow problem area (ILRI) while others have an extremely wide mandate (NBSS&LUP). They have a network of regional stations covering diverse agro-ecological areas to serve as a testing ground for the developed technologies. They are guided and serviced, both technically and administratively, by the ICAR headquarters. For ensuring greater effectiveness, from time to time, ICAR has taken steps to decentralize the decision-making process through delegation of powers to the institutes.

The individual institutes, under the supervision of a Director, are organized into well-defined divisions. On technical matters, they are assisted by scientific bodies like the Research Advisory Committee (RAC) represented by external experts, with Director as a member, and headed by an eminent person; and the Staff Research Council (SRC) which is represented by the institute scientists and headed by the Director. They meet once in a year and are charged with the responsibility to provide broad policy guidelines, and to plan, monitor, and evaluate research projects. All institutes have Management Committees except those with Deemed University status which have Management Boards. These Committees represented by the research and developmental personnel under the chairmanship of the Director, assist the institutes on broad policy formulation and financial matters. Once in five years, the performance of each institute is evaluated through a system of Achievement Audit Committees known as Quinquennial Review Teams. These teams, through a peer group review, play an important role in projecting the institutes' programmes in accordance with national policies and priorities. Besides, the Directors' Conferences, held regularly in the headquarters under the chairmanship of the Director General, consider problems common to the institutes.

Project Directorates: Because of the importance and magnitude of the work involved in a single commodity like rice, wheat and poultry, or a group of commodities like oilseeds, pulses and vegetables, ICAR has upgraded some of its research infrastructure/projects with added responsibilities, and designated them as Project Directorates. Except for the size and magnitude of work involved, these are basically the same as the Coordinated Research Projects. Additionally, they do undertake some research besides playing such national service roles like maintenance and supply of germplasm, organizing off-season nursery to promote and speed up research interests, monitoring pests and diseases, forecasting and issuing clearly warning about the pests and diseases outbreak, and performing such duties as a lead centre in relation to their respective subject matter, and so on. There are now ten of them under operation. They are:

1. Directorate of Rice Research, Hyderabad;
2. Directorate of Wheat Research, Karnal;
3. Directorate of Pulses Research, Kanpur;
4. Directorate of Oilseeds Research, Hyderabad;
5. Project Directorate on Vegetables, Varanasi;
6. Directorate of Cropping Systems Research, Modipuram;
7. Project Directorate on Water Management, Bhubaneswar;
8. Project Directorate on Cattle, Meerut;
9. Project Directorate on Poultry, Hyderabad;
10. Directorate of Maize Research, New Delhi.

National Research Centres: The National Commission on Agriculture recommended setting up of 'Centres of Fundamental Research' headed by eminent scientists in particular areas. Consequently, the ICAR conceived the idea of setting up a number of National Research Centres (NRCs). The concept of NRCs revolves around the need for concentrated attention with a mission approach by a team of scientists from different disciplines. They work under a senior leader on selected topics which have direct or indirect relevance to resolving national problems in a particular crop or commodity or a problem area of research. These centres are designed to concentrate on those crops and commodities not well served by the research institutes. Unlike the institutes, these centres do not have divisional set-up for individual disciplines nor have regional stations. They feed the national network of research with new materials, technology and information for subsequent adoption in the different production-oriented research programmes. The NRC for Groundnut was the first to be organized in 1979, and the NRC for Cashew is of the more recent ones established. There are now 31 such Centres, covering a wide range of areas like crops, horticulture, animal species, fisheries, resource management, etc. Five more approved centres are yet to be established. Some of the NRCs may grow into full-fledged institutes once their standard of work is established and if the subjects assume greater national importance.

Research Schemes/Projects: In addition to its institute-based research, ICAR promotes research schemes/projects in agriculture and allied areas to resolve location-specific problems. It is involved in a cooperative endeavor with other research organizations in carrying out multidisciplinary research programmes. Such promotional schemes fall under the following categories.

All India Coordinated Research Projects (AICRPs): These projects have been essentially conceived as an instrument to mobilize available scientific resources to find effective solutions for the national problems of agricultural production through inter-institutional interactions. The projects are developed as multidisciplinary and problem-oriented projects with major emphasis on multi-location testing of new materials/production systems. They provide opportunities for scientists working on similar problems in different institutions to come together, discuss and exchange ideas, information, and materials for mutual benefit. They also provide them with facilities for multi-location testing of improved technologies developed by various subsystems in different agro-climatic regions. The projects constitute an effective national grid of coordinated experiments by integrating different institutions and disciplines.

There are 70 such projects currently operating at 1,291 operating centres. Of these, 904 Centres are located in Agricultural Universities; 190 in ICAR Institutes; and the rest at other institutions. Crop sciences have 34 projects operating at 627 centres; and soil science and agricultural engineering have 20 projects at 364 centres. The expenditure on these projects has increased steadily and nearly one-fourth of the ICAR's budget is now spent on these projects. The complement of staff, determined on the basis of the technical work assigned and the nature of operation, are provided by the participating institutions, but paid for by the ICAR. The expenditure is shared by the ICAR and the collaborating institutions on 75:25 bases.

A high degree of accountability, based on continuous monitoring, is a noteworthy feature of these projects. Outstanding achievements have been made through these projects, and the development of such an approach has been a source of inspiration to many developing countries.

National Agricultural Research Project (NARP): Agricultural Universities which have a state-wide mandate for agriculture did not have a strong base for research at the regional level and most of the funds provided were utilized for developing the University main campuses, thus neglecting the regional research needs. To overcome this, the ICAR launched in 1979, with World Bank assistance, a novel scheme known as National Agricultural Research Project (NARP) to strengthen the regional research capabilities of these Universities for conducting need-based, location-specific and production-oriented research in identified agro-climatic zones. Under this project, each State is divided into a contiguous set of agro-climatic zones on the basis of climate, soils, crops and ecology. There are 131 such zones in the country with 125 zones in 17 States and Union Territories, and 6 zones in the North Eastern Region. In each zone, a major regional research station with a multidisciplinary team of scientists is established or strengthened.

The strengthening is accomplished by providing funds for incremental scientific and supporting staff, laboratory, equipment, transport, operating costs, and other physical facilities. The Agricultural University, on its part, has to provide funds for land, cultivation and station maintenance. The project also provides for strengthening the office of the Director of Research in these Universities for coordinating the research activities undertaken by these stations more effectively. An inventory of resources and systems of cultivation are prepared for each zone to serve as a basic document known as `Status Report' for identifying regional research needs. Farming systems research with emphasis on multidisciplinary approach to problem solving and establishment of a closer linkage between research and extension at the grassroots level with active participation of farmers are the essential components of this project. In the zonal workshops, organized at the beginning of each cropping season, research station scientists, developmental staff, and farmers' representatives review the progress and plan for the future based on actual needs. The project is guided and serviced by the NARP Directorate located in the ICAR headquarters. The progress of the project is regularly monitored by the Agricultural University and the ICAR, and is also reviewed periodically by the Special Mission of the World Bank. The ICAR's assistance is available for a period of five years after which the responsibility for continuing the project rests with the University.

The Project had the IDA support of $ 27 million in the first phase, which was about 50 per cent of the project cost. The first phase laid major emphasis on cereals, pulses and oilseeds under rainfed conditions. In this phase, nearly 109 sub projects were approved and it was closed in September 1985 after having used $19.5 million of the $ 27 million credit. The second phase on NARP, under implementation since February 1986, is intended to intensify and continue the process of decentralizing agricultural research. In addition to financing the sub-projects not completed under the first phase, this phase provides support for special support to regional research in the country.

2. Agricultural University System

As agriculture is a State subject, the responsibilities for research, education and extension rest with the State Governments. Prior to 1960, agricultural research in the States, essentially on local problems, was carried out by the State Departments of Agriculture supported by Agricultural Colleges. During the past 30 years, research and education have been transferred to the Agricultural Universities, and the State Departments of Agriculture organize extension services. The Universities are supported by their respective State Governments. ICAR provides financial support and assists their research and education programmes.

The University Education Commission (1949) recommended the setting up of `Rural Universities'. This was endorsed by the two Joint Indo-American Teams in 1955 and in 1959, as well as the Ford Foundation Study Team in 1959. In 1960, the Agricultural Universities Committee under the Chairmanship of Dr. Ralph W. Cummings prepared certain guidelines for the establishment of Agricultural Universities in different States, and the ICAR gave necessary support. The first Agricultural University was established at Pant Nagar in Uttar Pradesh (Nowadays in Uttarakhand) in 1960, patterned on the Land-Grant System of the United States. The Second Education Commission (1964-66) recommended at least one Agricultural University in each State, and ICAR prepared a Model Act in 1966. All the States have now at least one Agricultural University each. Though the Model Act specifies that only one University shall be established in each State, which was later endorsed by the National Commission on Agriculture, many States have established multiple Universities to meet regional needs. There are at present 67+ Agricultural Universities including the Central University in the North Eastern Region. These include two veterinary and animal science universities, one in Tamil Nadu and another in West Bengal. Some Agricultural Universities have affiliated colleges recommendations.

1. Acharya N. G. Ranga Agricultural University, Guntur
2. Dr. Y.S.R. Horticultural University, Venkataramannagudem, Tadepalligudem
3. Sri Venkateswara Veterinary University, Tirupati
4. Assam Agricultural University, Jorhat
5. Bihar Agricultural University, Bhagalpur

6. Bihar Animal Sciences University, Patna
7. Dr. Rajendra Prasad Central Agriculture University, Samastipur
8. Chhattisgarh Kamdhenu Vishwavidyalaya, Durg
9. Indira Gandhi Krishi Vishwavidyalaya, Raipur
10. Anand Agricultural University, Anand
11. Junagadh Agricultural University, Junagadh
12. Kamdhenu University, Gandhinagar
13. Navsari Agricultural University, Navsari
14. Sardarkrushinagar Dantiwada Agricultural University, Banaskantha
15. Chaudhary Charan Singh Haryana Agricultural University, Hisar
16. Lala Lajpat Rai University of Veterinary and Animal Sciences, Hisar
17. Maharana Pratap Horticultural University, Karnal
18. National Dairy Research Institute, Karnal
19. Chaudhary Sarwan Kumar Himachal Pradesh Krishi Vishvavidyalaya, Palampur
20. Dr. Yashwant Singh Parmar University of Horticulture and Forestry, Solan
21. Sher-e-Kashmir University of Agricultural Sciences and Technology of Jammu, Jammu
22. Sher-e-Kashmir University of Agricultural Sciences and Technology of Kashmir, Srinagar
23. Birsa Agricultural University, Kanke
24. Karnataka Veterinary, Animal and Fisheries Sciences University, Bidar
25. University of Agricultural and Horticultural Sciences, Shimoga
26. University of Agricultural Sciences, Bangalore
27. University of Agricultural Sciences, Dharwad
28. University of Agricultural Sciences, Raichur
29. University of Horticultural Sciences, Bagalkot
30. Kerala Agricultural University, Vellanikkara, Thrissur
31. Kerala University of Fisheries and Ocean Studies, Kochi
32. Kerala Veterinary and Animal Sciences University, Wayanad

33. Central Institute of Fisheries Education, Mumbai
34. Dr. Balasaheb Sawant Konkan Krishi Vidyapeeth, Dapoli
35. Dr. Panjabrao Deshmukh Krishi Vidyapeeth, Akola
36. Maharashtra Animal and Fishery Sciences University, Nagpur
37. Mahatma Phule Krishi Vidyapeeth, Rahuri
38. Vasantrao Naik Marathwada Krishi Vidyapeeth, Parbhani
39. Central Agricultural University, Imphal
40. Odisha University of Agriculture and Technology, Bhubaneswar
41. Guru Angad Dev Veterinary and Animal Sciences University, Ludhiana
42. Punjab Agricultural University, Ludhiana
43. Agriculture University, Jodhpur, Jodhpur
44. Agriculture University, Kota, Kota
45. Maharana Pratap University of Agriculture and Technology, Udaipur
46. Rajasthan University of Veterinary and Animal Sciences, Bikaner
47. Sri Karan Narendra Agriculture University, Jobner
48. Swami Keshwanand Rajasthan Agricultural University, Bikaner
49. Tamil Nadu Agricultural University, Coimbatore
50. Tamil Nadu Fisheries University, Nagapattinam
51. Tamil Nadu Veterinary and Animal Science University, Madhavaram, Chennai
52. P.V. Narasimha Rao Telangana Veterinary University, Hyderabad
53. Professor Jayashankar Telangana State Agricultural University, Hyderabad
54. Sri Konda Laxman Telangana State Horticultural University, Hyderabad
55. Banda University of Agriculture and Technology, Banda
56. Chandra Shekhar Azad University of Agriculture and Technology, Kanpur
57. Indian Veterinary Research Institute, Bareilly
58. Narendra Deva University of Agriculture and Technology, Faizabad
59. Rani Lakshmi Bai Central Agricultural University, Jhansi

60. Sam Higginbottom University of Agriculture, Techn & Sciences, Allahabad
61. Sardar Vallabhbhai Patel University of Agriculture and Technology, Meerut
62. U.P. Pt. Deen Dayal Upadhyay Pashu Chikitsa Vigyan Vishwavidyalaya Evam Go-Ansundhan Sansthan, Mathura
63. G. B. Pant University of Agriculture and Technology, Pantnagar
64. Veer Chandra Singh Garhwali Uttarakhand University of Horticulture & Forestry, Pauri Garhwal
65. Bidhan Chandra Krishi Viswavidyalaya, Mohanpur
66. Uttar Banga Krishi Viswavidyalaya, Cooch Behar
67. West Bengal University of Animal and Fishery Sciences, Kolkata

Deemed University

1. Indian Agricultural Research Institute, Pusa-110012, New Delhi
2. Indian Veterinary Research Institute, Izatnagar, Bareilly-243122, Uttar Pradesh
3. National Dairy Research Institute, Karnal-132001, Haryana
4. Central Institute of Fisheries Education, Mumbai-400061, Maharashtra
5. Allahabad Agricultural Institute, Allahabad-211007, Uttar Pradesh

Central University with Agriculture (as a subject)

1. Banaras Hindu University, Varanasi, U.P.
2. Aligarh Muslim University, Aligarh, U.P.
3. Vishwa Bharti, Shantiniketan, West Bengal
4. Nagaland University, Medizipherma, Nagaland

Research Setup in India

1. Research Infrastructure: Basically, the research infrastructure consists of an experiment station at the main campus and a number of research stations and substations located in different parts of the State. There are numbers of research stations belonging SAUs, working on location-specific problems. Generally, the research programmes are headed by the Directors of Research, who are assisted by the Associate Directors of Research located at the regional research stations within the State. Some Agricultural Universities have established Advanced Centres by combining related subjects in areas such as plant protection, genetics and plant breeding, agricultural engineering, agricultural economics, water technology, etc. In order to undertake need-based and location-specific research, networks of Zonal Research Centres have been established since 1979 with assistance from the World Bank under NARP. These Centres numbering about 131 in the country, each located in a distinct agro-climatic zone, is a part of the Agricultural Universities System.

2. Research planning, monitoring and evaluation: Agricultural Universities have State-wide responsibility for research in agriculture. In those States where, more than one University is there, the research responsibilities are shared on the regional basis. To ensure relevant research planning, their efficient implementation and proper evaluation, each Agricultural University has a Research Council or a Research Advisory Committee as an apex body for policy formulation and coordination of research activities. This body, chaired by the Vice-Chancellor comprises Director of Research, Director of Extension Education, Deans of constituent colleges, representatives of State Departments and farmers. It reviews periodically the overall status of research activities in the University, and determines their priorities and future direction. Research is organized under; (i) University research; (ii) Postgraduate student research; and (iii) Coordinated research programmes. The Director of Research, who is the overall in-charge of research, prepares an annual plan indicating the main thrusts of research within the broad directions given by the Research Advisory Committee. The Directorate of Research is responsible for research review and evaluation, and timely publication of research results and reports. By and large, the individual scientists formulate research projects which are then scrutinized by the concerned Department Head, examined at the Faculty/Departmental level and finally approved by the University Research Advisory Committee. Thereafter, the Director of Research and Department Heads provide funds and facilities to the scientists. In respect of research done by teachers and postgraduate students, the Director of Research acts in coordination with the Deans/Principals of the respective colleges.

Special mechanisms exist for the planning, monitoring, and evaluation of ICAR supported programmes in the Agricultural Universities. In the case of Coordinated Projects, the University scientists work in close co-operation with others from the ICAR Institutes and other Universities through the mechanism of All India Workshops conducted periodically. Ad hoc research schemes formulated by the University scientists are first scrutinized by the Scientific Panels and approved by the ICAR. These schemes are continuously monitored and evaluated by the ICAR in collaboration with the University. In the case of NARP, the programmes formulated at Zonal Research and Extension Advisory Committee meetings are constantly monitored and reviewed by the University concerned, the ICAR, and the World Bank Missions.

In most Agricultural Universities, the Research Evaluation Committees attended by the scientists and extension subject matter specialists provide the much needed in-house review mechanism within the Universities to examine the findings and data support emerging from various research projects. Only when the results are substantiated from trials, both on the experimental farms and farmers' fields, recommendations are made for large-scale implementation.

3. Other agencies imparting agricultural education

A. General Universities: Many General Universities with well-developed faculties in agriculture, or strong departments engaged in areas such as genetics, plant physiology, mycology, entomology, biochemistry, economics, chemistry, marine biology, home science, etc. have made distinctive contributions to agricultural research in the country. Besides, the Central Universities like the Banaras Hindu University, Santiniketan, etc. have Institutes/Schools of Agricultural Sciences which are engaged in research in agriculture and allied areas, some of which are supported by the ICAR.

B. Scientific Organizations: Many other scientific organizations either directly undertake research, or sponsor and support programmes related to agriculture. The Council of Scientific and Industrial Research (CSIR), through its network of National Laboratories, provides research support in areas like processing of agricultural products, recycling of agricultural wastes, development of various agro-chemicals, etc. The Indian Council of Medical Research's (ICMR) research on the nutritional qualities of various agricultural produce including toxicity and occupational health of agricultural workers have greatly helped the ICAR in planning its research programmes. Some of the areas in which the Bhabha Atomic Research Centre (BARC) is actively engaged are the development of newer varieties of crops and preservation of agricultural produce. The Indian Space Research Organization (ISRO) is helping the research system to assess India's soil and water resources.

Technological institutions like IIT, Kharagpur, are active in the fields of agricultural engineering, soil and water management, and agronomy. The Department of Science and Technology (DST) promotes research on genetic engineering, post harvest technology, and areas of basic sciences supportive to agriculture. The Department of Non-Conventional Energy Sources works on the utilization of solar and wind energies and biogas for agricultural purposes. The Department of Meteorology is actively engaged in research on crop-weather forecasting. The Department of Ocean Development is involved in assessing the fishery resources in the country and promotes research in the area of fisheries.

In addition, institutions like the National Dairy Development Board (NDDB) under the Agriculture Ministry; various Commodity Boards like Silk, Coffee, Rubber, Tea, and Cardamon Boards under the Commerce Ministry; and the Forest Research Institute and Wasteland Development Board under the Ministry of Forestry and Environment help in strengthening the agricultural research system in the country.

C. Private Sector: Involvement of private sector in agricultural research is of recent origin. In mid 1960s, several private companies started programmes mainly to develop hybrid maize, sorghum and bajra. Research on vegetables started in late 1960s. Private sector research is confined mainly to breeding crop hybrids, certain plantation crops, agro-chemicals, poultry, and agricultural machinery.

Private sector research in seed industry has grown very rapidly. Several private companies are now engaged in the production of hybrid seeds of a variety of crops like cotton, sorghum, bajra, maize, vegetables, red gram, rice, etc., and a small seed export industry has also emerged. There are at least ten private companies undertaking plant breeding research, and several others are involved in plant protection research. Besides their own research stations, these companies conduct experiments on farmers' fields. They test the bio-efficiency of insecticides and herbicides that are new to India as well as synthesize new compounds. Private research in poultry sector is of recent origin. Although Government introduced exotic birds in commercial poultry industry, they were popularized by the private sector and many companies are now actively engaged in it. Many other large industrial concerns are engaged in research on shrimps and shrimp feed.

Some private companies undertake major research and development programmes for the improvement of tractors and irrigation pumps. Research on tractors seems to be primarily aimed at improving quality, fuel efficiency and engine durability. Some companies are now moving into agricultural implements. Research in pump industry is aimed at increasing the efficiency of pumps through improved design and better materials. Some companies are even experimenting with non-conventional sources of power. Historically, private companies in the processing and plantation sector have been a very important source of new agricultural technology. Some of the prominent ones include Indian Sugar Mills Association, Southern Planters Association, Textile Mills Association, Silk Industry, etc. Some large firms are involved in research on animal nutrition, plant growth regulators, biotechnology like tissue culture in cardamom, sugarcane, coconut and tea, bio-fertilizers. Research in the area of tree farming including in vitro culture and tree breeding is also receiving attention of private firms.

Some of the well-established institutions such as Allahabad Agricultural Institute, Bharatiya Agro-Industries Foundation, Wool Research Association, and United Planters Association of South India undertake short-term, mission-oriented research projects supported by the ICAR for multi-locational testing of varieties and agro-techniques.

In order to promote scientific research and the participation of industry in it, the Government through the Income Tax Act of 1961 has offered certain tax concessions relating to the expenditure on scientific research. The involvement of private agencies in agricultural research is gaining momentum with greater sophistication in technological development and the prospects of high returns on investment in agriculture.

Technology Mission in Agriculture: In the Seventh Five Year Plan, a mission-oriented approach to technology development was emphasized to faster relevance and to provide motivation for establishing organic working linkages between different sectors, which otherwise remained compartmentalized. The Steering Group on Science and Technology, constituted by the National Planning Commission, has identified.

The Technology Mission on Oilseeds Research was set up in April 1986, to provide research and technology support to make the country self-reliant in edible and non-edible oils. The Mission concentrates its attention on major oilseed crops like groundnut, rapeseed, mustard, soybean, sunflower, safflower, linseed, sesame, and Niger. It also gives priority to non-edible oilseed crops to meet the requirements of industry. The Mission envisages an integrated approach involving different developmental, scientific, input, banking, and marketing agencies. A total of 180 districts are earmarked for the purpose. The Department of Agriculture and Cooperation in the Ministry of Agriculture and the ICAR are the principal implementing agencies. The Additional Secretary to the Government of India in the Department of Agriculture and Cooperation is the Mission Director. Four Sub-Missions share the operational responsibilities. These are: (a) Production Technology (R & D) Sub-Mission for which the Director General of ICAR is the Chairman; (b) Input Supply and Production Sub-Mission for which the Additional Secretary to the Government of India is the Chairman, and the Agricultural Commissioner is the Co-chairman; (c) Postharvest Technology and Processing Sub-Mission for which the Scientific Advisor to the Planning Commission is the Chairman; and (d) Pricing, Transport, Procurement and Marketing Sub-Mission for which the Additional Secretary to the Government of India is the Chairman. Quarterly reviews of all the developmental activities are done to effect mid-term corrections. An expert team was set up recently to review the progress. Because of its significant achievement within a short span of four years by increasing the production of oilseeds to over 20 million tones, it is decided to continue the Mission during the Eighth Plan Period.

Conclusion

Department of Agricultural Research and Education (DARE) coordinate and promote agricultural research and education in the country. DARE provides the all-important government linkages for the Indian Council of Agricultural Research (ICAR), the other research organization for co-coordinating, allegorical and managing research, education and Extension in agronomics including horticulture, fisheries and Animal sciences in the absolute country.

The setup has got 102 research institutes and 73 agricultural universities (at present, as on June, 2018) advance beyond the country, the National Agricultural Research System (NARS) is one of the better in the world. ICAR Institutes accept the Departments of Extension, the Agricultural Universities accept the directorates of Extension to identify, research and clarify the communicable agricultural technologies and added accompanying advice appearing from research as able-bodied as to authorize able communication a part of assorted departments and added stakeholders both the accessible and non-public for their transfer. KVKs as well accommodate Technical backstopping to 'Agriculture Technology Management Agencies' (ATMA), an arrangement adjourned by the Department of Agronomics Cooperation and Farmers Welfare (DAC & FW) and added agencies at the commune level. A commune Collective Action Plan for acceptable interface amid scientists, Extension functionaries and farmers is able through a collective affair of KVK and ATMA admiral under the chairmanship of the Commune Collector. An annual interface meeting, involving all band departments of the district, is captivated by all KVKs to advise the accomplishing of the Collective Action Plan and to allotment new advice and technologies for added broadcasting in the district.

The Indian Council of Agricultural Research as well maintains able linkages with added key programs of the Government of India like RKVY, MNREGA and NHM and added Extension activities including those accompanying to accessible sector, farmers' organizations, NGOs, Commodity Boards and non-public-sector. ICAR contributes through technology backstopping in the National *Rabi* and *Kharif* conferences organized every year in which Extension agencies from all the states as well participate.

Some pertinent short questions

1. Write full form of RKVY, MNREGA, NHM, NGO and NARS;
2. Write down national setup at ICAR system;
3. Write down national setup of SAU system;
4. How much deemed university we have?
5. What the role of NBPGR in modern agriculture?

References

http://beasiswascholarships.com/search/scholarship%20and%20fellowship%20for%20scientific%20mobility%20in%20india

http://beasiswascholarships.com/search/scholarship%20and%20fellowship%20for%20scientific%20mobility%20in%20india

16

Current Scenario and Journey of Indian Agriculture and Its Impacts

Introduction

Indian agriculture has come a long way since the inception of planning in1951. All along there was an almost obsessive concern of development policy with the attainment of self-sufficiency in food. We faced nightmare of absolute shortage of food grains supplies in the 1960s, when the average annual imports of wheat hovered around three million tones. We survived through this phase because of the munificence of the US, which supplied wheat to us under its PL 480 programme. There have been several policy statements for agriculture during the last sixty years. India attained self-sufficiency in food grains in the 1970s and what is more, has emerged as an exporter of food grains in more recent years. Attainment of food self-sufficiency, even in a technical sense, is admittedly an important landmark in the history of growth of Indian agriculture, but this has not meant the dilution of problems facing Indian agriculture. Problems of agricultural growth and rural development in general continue to remain, in the new millennium, as daunting as they were in the 1950s. One major difference is that agricultural sector today faces a host of what we might call second generation problems.

Recent Agricultural Production Scenario

Agriculture plays a vital role in India's economy. 54.6% of the population is engaged in agriculture and allied activities (census 2011) and it contributes 17% to the country's Gross Value Added (GVA, current price 2015-16, and 2011-12 series). Given the importance of agriculture sector, Government of India took several steps for its sustainable development. As per the land use statistics 2013-14, the total geographical area of the country is 328.7 million hectares, of which 141.4 million hectares is the reported net sown area and 200.9 million hectares is the gross cropped area with a cropping intensity of 142 %. The net sown area works out to be 43% of the total geographical area. The net irrigated area is 68.2 million hectares. There has been a continuous decline in the share of Agriculture and Allied sector in the GVA from 18.2 percent in 2012-13 to 17.0 percent in 2015-16 at current prices.

Table 1: Falling share of Agriculture and Allied sector in GVA is an expected outcome in a fast growing and structurally changing economy.

Share of Agriculture & Allied Sectors in Total GVA
(Rs. in Crore)

Items	Year			
	2012-13	2013-14	2014-15	2015-16
GVA of Agriculture and Allied Sectors	1680797	1902452	1995251	2093081
Per cent to total GVA	18.2	18.3	17.4	17.0

Source: Central Statistics Office, Ministry of Statistics and Programme Implementation, Govt. of India.

As of 2009-10, more than half of the total workforce of the country, i.e. 243 million persons were employed in agriculture. The share of population depending on agriculture for its livelihood consists of landowners, tenant farmers who cultivate a piece of land, and agricultural labourers who are employed on these farms. Agricultural output has been volatile over the past 10 years, with annual growth ranging from 8.6% in 2010-11, to -0.2% in 2014-15 and 0.8% in 2015-16. The agriculture sector's contribution to the Gross Domestic Product (GDP) decreased from 54% in 1950-51 to 15.4% in 2015-16, while that of the services sector increased from 30% to 53%. While the agriculture sector's contribution to GDP has decreased over the past few decades, the contribution of sectors such as manufacturing (employing 10.5% of the population) and services (employing 24.4% of the population) has increased.

Agricultural productivity depends on several factors. These include the availability and quality of agricultural inputs such as land, water, seeds and fertilizers, access to agricultural credit and crop insurance, assurance of remunerative prices for agricultural produce, and storage and marketing infrastructure, amongst others.

Notwithstanding the setback in *kharif* crops 2015-16 due to bad monsoon and Rabi crops due to warm winter, total food grain production in the country increased marginally in 2015-16. As per 4th Advance Estimates for 2015-16, total production of rice is estimated at 104.32 million tons which is lower by 1 .17 million tons than the production of 105.48 million tons during the preceding year. Production of wheat, estimated at 93.50 million tons, is higher by 6.97 million tons than the production of 86.53 million tons achieved during 2014-15. Total production of Coarse Cereals estimated at 37.94 million tons is lower by 4.38 million tons than their production during 2014-15. Total food grains production in the country is estimated at 252.22 million tons which is marginally higher by 0.20 million tons than the previous year's food grains production of 252.02 million tons. Production of pulses estimated at 16.47

million tons is lower by 0.68 million tons than their production during 2014-15. Total production of food grains increased from 51 million tons in 1950-51 to 252 million tons in 2015-16. According to the second advance estimate by the Ministry of Agriculture, food grains production is estimated to be 272 million tons in 2016-17. The production of wheat and rice took off after the green revolution in the 1960s, and as of 2015-16, wheat and rice accounted for 78% of the food grains production in the country.

With a decline of 2.21 million tons over the last year, total production of oilseeds in the country is estimated at 25.30 million tons. Production of sugarcane is estimated at 352.16 million ton. Total production of cotton estimated at 30.15 million bales (of 170 kgs each) is lower than previous year's production of 34.81 million bales and also lower by 4.48 million bales than its five years average production of 34.63 million bales. Production of jute & mesta estimated at 10.47 million bales (of 180 kg each) is lower by 0.66 million bales (6.3%) than their production during the previous year.

Table 2: Sector, production and yield of major Crops

Crops	Sector (Lakh hectare)			Production (Million Tons)			Yield (kg/hectare)		
	2013-14	2014-15	2015-16*	2013-14	2014-15	2015-16*	2013-14	2014-15	2015-16*
Rice	441.36	441.10	433.88	106.65	105.48	104.32	2416	2391	2404
Wheat	304.73	314.65	302.27	95.85	86.52	93.50	3145	2750	3093
Coarse cereals	252.19	251.70	237.75	43.29	42.86	37.93	1717	1703	1596
Pulses	252.12	235.54	252.59	19.25	17.15	16.47	764	728	652
Food grains	1250.41	1243.00	1226.50	265.04	252.02	252.22	2120	2028	2056
Oilseeds	280.50	255.96	261.34	32.74	27.51	25.30	1168	1075	968
Sugarcane	49.93	50.66	49.53	352.14	362.33	352.16	70522	71512	71095
Cotton@	119.60	128.19	118.72	35.90	34.80	30.15	510	462	432
Jute and Mesta#	8.38	8.09	7.85	11.69	11.12	10.47	2512	2473	2399

* 4th advance estimates; @ production in million bales of 170 kg each; # production in million bales 180 Kg. each.

Table 3: Area, production and yield of major Crops

Crops	Area (Lakh hectare)			Production (Million Tons)			Yield (kg/hectare)		
	2013-14	2014-15	2015-16*	2013-14	2014-15	2015-16*	2013-14	2014-15	2015-16*
Rice	441.36	441.10	433.88	106.65	105.48	104.32	2416	2391	2404
Wheat	304.73	314.65	302.27	95.85	86.52	93.50	3145	2750	3093
Coarse cereals	252.19	251.70	237.75	43.29	42.86	37.93	1717	1703	1596
Pulses	252.12	235.54	252.59	19.25	17.15	16.47	764	728	652
Food grains	1250.41	1243.00	1226.50	265.04	252.02	252.22	2120	2028	2056
Oilseeds	280.50	255.96	261.34	32.74	27.51	25.30	1168	1075	968
Sugarcane	49.93	50.66	49.53	352.14	362.33	352.16	70522	71512	71095
Cotton@	119.60	128.19	118.72	35.90	34.80	30.15	510	462	432
Jute & Mesta#	8.38	8.09	7.85	11.69	11.12	10.47	2512	2473	2399

* 4th advance estimates; @ Production in million bales of 170 kg each; # Production in million bales 180 Kg. each.

Capital Formation in Agriculture and Allied Sectors

Gross Capital Formation (GCF) in Agriculture and Allied sector is estimated separately for public, private corporate and the household sectors. Items included in the estimates of GCF are:

(i) Improvement of land and Irrigation works

(ii) Laying of new orchards and plantations

(iii) Purchase of agricultural machinery and implements

(iv) Agricultural construction works

(v) Additions to livestock

(vi) Fishing boats and nets etc.

Gross Capital Formation (GCF) in Agriculture and Allied sector relative to GVA in this sector has been showing a fluctuating trend from 18.3 per cent in 2011-12 to 16.2 per cent in 2014-15.

Table: 3 Gross Capital Formation (GCF) in Agriculture and Allied sectors relative to Gross Value Added (GVA) at 2011-12 basic prices

(Rs. in crore)

Period	GCF in Agriculture & Allied Sectors			GVA in Agriculture & Allied Sectors	GCF as percentage of GVA		
	Public	Private	Total		Public	Private	Total
2011-12	35715	238717	274432	1501816	2.4	15.9	18.3
2012-13	36078	217230	253308	1524398	2.4	14.3	16.6
2013-14	32472	244694	277165	1588237	2.0	15.4	17.5
2014-15 *	36061	220434	256495	1584293	2.3	13.9	16.2

Source: Central Statistics Office, MOSPI

*As per Advance Estimates for 2015-16 (latest available) released on 8.2.2016

National Policy for Farmers (NPF), 2007

Government of India approved the National Policy for Farmers (NPF) in 2007. The Policy provisions, inter alia, include asset reforms in respect of land, water, livestock, fisheries, and bio-resources; support services and inputs like application of frontier technologies; agricultural bio-security systems; supply of good quality seeds and disease-free planting material, improving soil fertility and health, and integrated pest management systems; support services for women like crèches, child care centres, nutrition, health and training; timely, adequate, and easy reach of institutional credit at reasonable interest rates, and farmer-friendly insurance instruments; use of Information and Communication Technology (ICT) and setting up of farmers' schools to revitalize agricultural extension; effective implementation of MSP across the country, development of agricultural market infrastructure, and rural non-farm employment initiatives for farm households; integrated approach for rural energy and others.

Many of the provisions of the NPF are being operationalized through various schemes and programmes which are being implemented by different Central Government Departments and Ministries. For the operationalization of the remaining provisions of the Policy, an Action Plan has been finalized and circulated to the Ministries and Department concerned, as well as to all States and UTs for necessary follow up action. An Inter-Ministerial Committee constituted for the purpose also monitors the progress of the Plan of Action for the operationalization of the NPF.

Doubling the farmer's income

To achieve the target of doubling of income of farmers by 2022, as announced during presentation of Budget, 2016-17, a Committee has been constituted under the Chairmanship of Additional Secretary, Department of Agriculture, Cooperation and Farmers Welfare to examine the various issues involved. All the State Governments have been requested to hold across-the–board deliberations and prepare strategy for the states that will lead to doubling the income of the farmers. NABARD has also organized six Regional Conferences to sensitize all the States for the strategy that will lead to doubling the income of farmers by 2022.

The strategy of the Government is to focus on farmers' welfare by making farming viable. Farm viability is possible, when cost of cultivation is reduced, yield per unit of farm is increased and farmers get remunerative prices on their produce. The department of agriculture, government of India is implementing various schemes to meet this objective with a mission mode approach. In this chapter some important missions, their salient features, fond allocation and progress so far have been highlighted.

Rashtriya Krishi Vikas Yojana (RKVY)

The beginning of **Rashtriya Krishi Vikas Yojana (RKVY)** is having background. National Development Council (NDC), in its meeting held on 29th May, 2007 resolved that a special Additional Central Assistance (ACA) Scheme be introduced to incentivize States to draw up comprehensive agriculture development plans, taking into account agro-climatic conditions, natural resources and technology for ensuring more inclusive and integrated development of agriculture and allied sectors. Accordingly, Department of

Agriculture & Cooperation (DAC), Ministry of Agriculture, in consultation with the Planning Commission, launched Rashtriya Krishi Vikas Yojana in 2007-2008.

During XI Plan, Rs. 22,408.77 crore was released for implementation of the scheme in the States & Union Territories (UTs) of which Rs. 22,332.97 crore was utilized in implementing projects relating to crop development, horticulture, agricultural mechanization, natural resource management, marketing and postharvest management, animal husbandry, dairy development, fisheries, extension etc. During XII Five Year Plan, an outlay of Rs. 63,246 crore has been earmarked for implementing RKVY, out of which Rs. 8400 crore, Rs. 7052.51crore, Rs. 8443.20 crore and Rs.3942.38 crore had been released under the scheme during 2012-13, 2013-14, 2014-15 and 2015-16 respectively.

Based on the feedback received from States, experiences garnered during implementation in 11th Plan and inputs provided by Stakeholders, Operational Guidelines of RKVY have been revised for implementation from 2014-15 in order to enhance efficiency and efficacy of the programme and its inclusiveness during 12th Plan period. The revised Operational Guidelines of the scheme mandated that at least 35% of Normal RKVY allocation should be utilized by the States for implementing Infrastructure and Assets development projects.

Objectives of RKVY Project

i. To incentivize the States so as to increase public investment in agriculture and allied sectors.

ii. To provide flexibility and autonomy to States in the process of planning and executing agriculture and allied sector schemes.

iii. To ensure the preparation of Agriculture Plans for the districts and the States based on agro-climatic conditions, availability of technology and natural resources.

iv. To ensure that the local needs/crops/priorities are better reflected in the agricultural plans of the States.

v. To achieve the goal of reducing the yield gaps in important crops through focused interventions.

vi. To maximize returns to the farmers in agriculture and allied sectors;

vii. To bring about quantifiable changes in the production and productivity of various components of agriculture and allied sectors by addressing them in a holistic manner.

Sub-Schemes launched under RKVY

RKVY has enabled launching of new schemes/programmes keeping States' flexibility and authority intact. Since 2010-11 several sub-schemes have been introduced under RKVY with focused objectives. Following special Programmes/schemes are being implemented as sub-schemes of RKVY, which also include some new sub- schemes launched during the current financial year, with a total allocation of Rs.1069.20 crore for 2016-17.

i. **Bringing Green Revolution to Eastern India (BGREI)** which has been discussed earlier.

ii. **Crop Diversification Programme (CDP)** which has already been discussed.

iii. **Reclamation of Problem Soil:** This programme is initiated as a sub-scheme of RKVY during 2016-17 for enhancing soil fertility & productivity for meeting demand of food grain of the country. The allocation for the scheme for the current financial year is Rs.50 crore.

iv. **Rice Fallow Areas in Eastern India for Pulses and Oil Seeds:** This is a new initiative launched under RKVY as a sub-scheme to bring area of rice fallow in Eastern India under pulses and oilseed cultivation. The allocation for the scheme for 2015-16 was Rs.50 crore.

v. **Foot and Mouth Disease (FMD) Control Programme**: This sub-scheme of RKVY is initiated during 2016-17 for supplementing the effort of the State Governments for prevention, control and containment of FMD. Rs.100.65 crore was allotted for the scheme for 2015-16.

A well designed implementation strategy has been made to materialize RKVY and it accords flexibility and autonomy to States in planning and executing projects related to agriculture and allied sectors. States are empowered to formulate strategies for development of the agriculture and allied sectors in a holistic way taking into account their agro-climatic conditions so as to effectively address their local needs and priorities, rather than continuing with business as usual mode of "one size fits all" formula.

Besides planning and execution of projects, RKVY provides complete flexibility and autonomy to States to customize interventions as per local requirements. For a vast country like India with diverse soil and agro-climatic conditions, varying technological capabilities and agronomic practices, RKVY framework has enabled States to plan and chart their own developmental trajectory. At national level, it enables the country to achieve the desired annual growth rate in agriculture and allied sectors.

Over the years, RKVY has also facilitated in addressing national priorities, without affecting the autonomy and flexibility of States, through special programmes as sub-schemes. Towards this endeavour, sub-schemes such as Bringing Green Revolution to Eastern India (BGREI), Crop Diversification Programme (CDP), Reclamation of Problem Soil, Foot and Mouth Diseases Control Programme, Rice Fallow Areas in Eastern India for Pulses and Oil Seeds etc are being implemented during 2016-17.

RKVY is administered by the Union Ministry of Agriculture & Farmers Welfare. Funds under this scheme are provided to the States as grant-in-aid by the Central Government.

RKVY performances

(i) Bringing Second Green Revolution to Eastern India (BGREI)

Bringing Green Revolution to Eastern India (BGREI), a sub scheme of Rashtriya Krishi Vikas Yojana was initiated in 2010-11 to address the constraints limiting the productivity of "rice based cropping systems" in eastern India comprising seven states namely, Assam, Bihar, Chhattisgarh, Jharkhand, Odisha, Eastern Uttar Pradesh and West Bengal.

An amount of Rs. 400 crores each was allocated for the programme during 2010-11 & 2011-12 and an amount of Rs. 1000 crores each during 2012-13 to 2014-15. Till 2014-15 the programme was implemented on 100 % GOI assistance. However, from the year 2015-16, the programme is being implemented on 60:40 sharing basis between GOI and States and 90:10 sharing basis in NE State (Assam). An amount of Rs. 630.00 crore as central share has been earmarked for implementation of the programme during 2016-17.

Under this programme, interventions like (i) block demonstrations of rice and wheat in cluster mode; (ii) asset building activities such as construction of shallow tube, Dug well / Bore well, promotion of farm implements like seed drill, pump set, cono-weeder, drum seeder and other need based farm implements for cultivation of Rice/ wheat as per NFSM-Rice/ Wheat; (iii) need based site specific activities and (iv) marketing support are also included.

(ii) Crop Diversification Programme

Crop Diversification Programme is being implemented in Original Green Revolution States viz: Punjab, Haryana and Western Uttar Pradesh **as a sub scheme of RKVY** since 2013-14 to divert the area of water guzzling paddy to alternate crops like pulses, oilseeds, maize, cotton and agro forestry with the objective of tackling the problem of declining of soil fertility and depleting water table in these states. In order to encourage tobacco growing farmers to shift to alternate crops/cropping systems, Crop Diversification Programme has been extended to tobacco growing states of Andhra Pradesh, Bihar, Gujarat, Karnataka, Maharashtra, Odisha, Tamil Nadu, Telangana, Uttar Pradesh and West Bengal w.e.f. 2015-16.

An amount of Rs. 500.00 crore & Rs. 250.00 crore were allocated for Crop Diversification Programme in previous Green Revolution States (Punjab, Haryana and Western Uttar Pradesh) during year 2013-14 & 2014-15 on 100 % GOI assistance. However, from the year 2015-16, the programme is being implemented on 60:40 sharing basis between GOI and State Governments. Financial allocation of Rs. 150.00 crore as a central share (Rs. 125.00 crore for Crop Diversification Programme in Original Green Revolution States and Rs. 25.00 crore for replacing tobacco farming with alternate crops/cropping system) have been allocated during 2015-16. An amount of Rs. 180.00 crore as a central share (Rs. 150.00 crore for Crop Diversification Programme in Original Green Revolution States and Rs. 30.00 crore for replacing tobacco farming with alternate crops/cropping system have been earmarked during 2016-17.

National Food Security Mission (NFSM)

National Food Security Mission was launched in 2007-08 to increase the production of rice, wheat and pulses by 10, 8 and 2 million tons, respectively by the end of 11th Plan through area expansion and productivity enhancement; restoring soil fertility and productivity; creating employment opportunities; and enhancing farm level economy. The Mission has continued during 12th Plan with new target of additional production of 25 million tons by the end of 12th Plan and promotion of commercial crops like cotton, jute and sugarcane.

The basic strategy of the Mission is to promote and extend improved technologies, i.e., seed, micro-nutrients, soil amendments, integrated pest management, farm machinery and implements, irrigation devices along with capacity building of farmers. The major interventions/activities covered under NFSM include cluster demonstrations of rice, wheat, pulses and coarse cereals, distribution of improved varieties/hybrid seeds, need based inputs, resource conservation techniques / energy management, efficient water/ application tools, cropping system based trainings and local initiatives; award for best performing districts etc.

NFSM is implemented in 638 districts of 29 states of which NFSM-Rice is implemented in 194 districts of 25 states during 12th Plan in 2016-17; NFSM-Wheat in 126 districts of 11 states; NFSM- Pulses in 638 districts of all 29 States; NFSM-Coarse cereals in 265 districts of 28 States.

A three-tier monitoring mechanism was inbuilt in NFSM at National, State and District levels. At National level General Council (GC) under the chairmanship of Union Agriculture Minister and Farmers Welfare was constituted to oversee the implementation of NFSM and take policy decisions for mid-term corrections. Altogether, there are eight (8) National Level Monitoring Teams (NLMTs). At state level, the monitoring is undertaken by State Food Security Mission Executive Committee (SFSMEC) under the Chairmanship of Chief Secretary. At district level, the monitoring is undertaken by District Food Security Mission Executive Committee (DFSMEC) under the chairmanship of District Collector.

Outcome of NFSM

The focused and target oriented implementation of mission initiatives has resulted in bumper production of wheat, rice and pulses in the country. During 2014-15, the productions of rice, wheat and pulses have been achieved at the level of 105.48 million tons, 86.53 million tons and 17.15 million tones, respectively As per 4th Advance Estimates during 2015-16, the production of wheat increased from 75.81 million tons in 2006-07 (pre-NFSM year) to 93.50 million tons, i.e. an increase of 17.69 million tons. Similarly, the total production of rice increased from 93.36 million tons in 2006-07 to 104.32 million tons in 2015-16 i.e. an increase of nearly 10.96 million tons. The total production of pulses also increased from 14.20 million tons during 2006-07 to 16.47 million tons during 2015-16 with an increase of 2.27 million tons.

National Mission on Oilseeds and Oil Palm (NMOOP)

Vegetable Oil constitutes an important part of our daily diet being source of energy, essential fatty acids and amino acids. Domestic consumption of edible oils has increased substantially over the years and has touched the level of 24.16 million tons in 2014-15 (Nov-Oct) and is likely to increase further with enhancement in income and population against the domestic availability of 9.58 million tons. Considering the importance of oilseeds, various oilseeds development schemes have been funded by the Government for encouraging cultivation of oilseeds including Oil Palm. The diverse agro-ecological conditions in the country are favorable for growing nine annual oilseeds which include seven edible oilseeds viz. groundnut, rapeseed-mustard, soybean, sunflower, sesame, safflower and niger and two non-edible oilseeds, castor and linseed.

The Centrally Sponsored Scheme of Integrated Scheme of Oilseeds, Oil Palm and Maize (ISOPOM), which was launched in 2004-05 and remained under implementation till March, 2014, Oil Palm Area Expansion (OPAE) programme, a sub-scheme of RKVY implemented during 2011-12 to 2013-14 have made significant contribution in increasing the oilseeds production and area expansion under oil palm. Implementations of these schemes have given fillip in augmenting the availability of vegetable oil in the country.

The scheme of ISOPOM, Tree Borne Oilseeds (TBOs) and Oil Palm Area Expansion (OPAE) programme have been restructured into National Mission on Oilseeds and Oil Palm (NMOOP) during 12th Plan and launched during 2014-15 and are implemented in 28 States. The sharing pattern of NMOOP between GOI and States is 60:40 for general states and 90:10 for NE and hilly states. Few components especially seed production & distribution, minikit of oilseeds for farmers, research and development (R&D) support, Front Line Demonstrations (FLDs) by Krishi Vigyan Kendras (KVKs) and other ICAR institutes are on 100% funding for Central Agencies and ICAR/SAUs.

Oilseeds are raised mostly under rainfed conditions and important for the livelihood of small and marginal farmers in arid and semi-arid areas of the country. The production of oilseeds has increased from 24.35 million tons in 2004-05 to 27.51million tons in 2014-15 and 25.30 million tons in 2015-16. The yield of oilseeds was 885 kg per hectare in 2004-05, increased to 1037 kg per hectare in 2014-15 and 968 kg per hectare in 2015-16 (as per 4th Advance Estimates). The highest production and yield of oilseeds were achieved during the year 2013-14 i.e. 32.75 million tons and 1167 kg per year, respectively. During 2014-15 and 2015-16, major oilseeds producing states experienced late monsoon at the time of sowing, insufficient rain during crop growing phase, untimely rain during pod maturity stage, yellow mosaic virus in soybean. As per the 1st Advance estimate of *Kharif* 2016- 17, oilseeds production is 23.36 million tones which are much higher than *Kharif* 2014-15 and 2015-16. The area, production and yield of oilseeds during last three years are as under:

Table 4: Recent Oilseeds production scenario of India

(Area in million hectare, Production in million tons & Yield in Kg/ha)

Year	Kharif			Rabi			Total		
	Area	Prod.	Yield	Area	Prod.	Yield	Area	Prod.	Yield
2004-05	17.24	14.15	820	10.28	10.20	993	27.52	24.35	885
2014-15	18.21	19.22	1056	7.39	8.32	1130	25.73	27.51	1037
2015-16*	18.85	16.59	880	7.29	8.71	1195	26.13	25.30	968
2016-17@	18.75	23.36	1246						

*As per 4th Advance Estimates; @As per 1st Advance Estimates (only Kharif).

The crop wise average production of nine major oilseeds during the 12th plan period and for the year 2012-13 to 2016-17are as follow:

Table 5: Crop wise average productions of nine major oilseeds during the 12th plan period and for the year 2012-13 to 2016-17

(Production in million tons)

Crop	Average of XI Plan (2007-08 to 2011-12)	2012-13	2013-14	2014-15	2015-16	2016-17 @
Groundnut	7.40	4.70	9.71	7.40	6.77	6.50
Castor	1.38	1.96	1.73	1.87	1.65	1.73
Sesame	0.74	0.69	0.72	0.83	0.87	0.68
Niger	0.11	0.10	0.10	0.08	0.08	0.10
Rapeseed & Mustard	6.89	8.03	7.88	6.28	6.82	0.00
Linseed	0.16	0.15	0.14	0.16	0.13	0.00
Safflower	0.18	0.11	0.11	0.09	0.06	0.00
Sunflower	0.93	0.54	0.50	0.43	0.33	0.13
Soybean	11.16	14.67	11.86	10.37	8.59	14.22
Total Nine Oilseeds	28.93	30.94	32.75	27.51	25.30	23.36

@As per 1st Advance Estimates (only Kharif).

The strategy to implement the Mission includes increasing Seed Replacement Ratio (SRR) with focus on varietal replacement; increasing irrigation coverage under oilseeds, diversification of area from low yielding cereals crops to oilseeds crops; inter-cropping of oilseeds with cereals/pulses/sugarcane; use of fallow land after paddy/potato cultivation; expansion of cultivation of Oil Palm, increasing availability of quality planting materials of Oil Palm & TBOs maintenance cost and inter-cropping during gestation period of oil palm and TBOs would provide economic return to the farmers when there is no production. The scheme would be implemented in a mission mode through active involvement of all the stakeholders. Fund flow would be monitored to ensure that benefit of the Mission reaches the targeted beneficiaries in time to achieve the targeted results. The NMOOP is further divided into following three segments for the specific targets.

Mini Mission I on Oilseeds

The target of the mini-mission-I is to achieve production of 35.51 million tons and productivity of 1328 kg/ha of oilseeds from the present average production & productivity of 28.93 million tons and 1081 kg/ha during the 11th Plan period respectively.

Mini Mission-II on Oil Palm

The target of the mini-mission-II is to bring additional 1.25 lakh hectare area under oil palm cultivation through area expansion approach in the states and maximization of productivity of fresh fruits bunches (FFBs) from 4927 kg per ha to 15000 kg per ha.

Mini Mission-III on TBOs (Tree borne Oilseeds)

The target of this mini-mission is to enhance seed collection of TBOs.

In order to bring more area under oil palm cultivation, Government of India has launched a new National Mission on Oilseeds and Oil Palm (NMOOP) during the 12th Five Year Plan under which Mini Mission-II is dedicated to oil palm area expansion, is being implemented in 12 states during 2016- 17. Under the Mission, financial assistance are being provided to the farmers @ 85% cost of the planting material and @ 50% cost of the other components like maintenance cost of new plantations for four years, installation of drip-irrigation systems, diesel/electric pump sets, bore-well/water harvesting structures/ponds, inputs for inter-cropping in oil palm (during gestation period), construction of vermi-compost units and purchasing of machinery & tools etc.

Moreover, 100% support is being provided to Indian Institute of Oil Palm Research (IIOPR), Pedavegi, Andhra Pradesh for Research & Development on oil palm. All these developmental efforts have resulted in area expansion under oil palm from 8585 ha in 1991-92 to 3.00 lakh ha by the end of 2015-16. Similarly, the Fresh Fruit Bunches (FFBs) production and Crude Palm Oil (CPO) have increased from 21,233 MT and 1,134 MT respectively (1992-93) to 12.83 lakh and 2.17 lakh MT respectively during the year 2015-16.

National Mission for Sustainable Agriculture (NMSA)

National Mission for Sustainable Agriculture (NMSA) is one of the eight Missions outlined under National Action Plan on Climate Change (NAPCC). The Mission aims at promoting sustainable agriculture through seventeen deliverables focusing on ten key dimensions of Indian Agriculture. During 12th Five Year Plan, these dimensions have been embedded and mainstreamed into Missions/Progammes/Schemes of Department of Agriculture, Cooperation & Farmers Welfare (DAC&FW) through a process of restructuring and convergence. NMSA as a programmatic intervention made operational from the year 2014-15 aims at making agriculture more productive, sustainable, and remunerative and climate resilient by promoting location specific integrated/ composite farming systems; soil and moisture conservation measures; comprehensive soil health management; efficient water management practices and mainstreaming rainfed technologies.

A dedicated website of National Mission for Sustainable Agriculture (NMSA) has been operationalized (http://nmsa. dac.gov. in/) to keep all information of the scheme in the public domain for access and use of all stakeholders.

The major components of NMSA are:

Rainfed Area Development (RAD)

RAD focuses on Integrated Farming System (IFS) for enhancing productivity and minimizing risks associated with climatic variability. Under this system, crops/cropping system is integrated with activities like horticulture, livestock, fishery, agro-forestry, apiculture etc. to enable farmers not only in maximizing farm returns for sustaining livelihood, but also to mitigate the impacts of drought, flood or other extreme weather events with the income opportunity from allied activities during crop damage. During 2016-17, RAD is being implemented in 27 States of the Country.

Integrated Farming System is being promoted under RAD in which activities like horticulture, livestock, fishery, agroforestry, value addition are to be taken up along with crops/cropping system. As on 31.12.2016, IFS activities of about 32740 ha have been covered comprising 4245 ha of livestock based farming system, 9727 ha of dairy based farming system, 1609 ha of fishery based farming system, 7408 ha of horticulture based farming system, 5105 ha of agro-forestry based farming system, 472 ha of silvi-pasture based farming system & 4174 ha of cropping system with peripheral plantations. Rs.157.14 crore has been released to the implementing states till 31st December, 2016.

Soil Health Management (SHM)

SHM is aimed at promoting location as well as crop specific sustainable soil health management including residue management, organic farming practices by way of creating and linking soil fertility maps with macro-micro nutrient management, appropriate land use based on land type.

Soil Health Management (SHM) is one of the most important interventions under National Mission for Sustainable Agriculture (NMSA). SHM also aims at promoting Integrated Nutrient Management (INM) through judicious use of chemical fertilizers including secondary and micro nutrients in conjunction with organic manures and bio-fertilizers for improving soil health and its productivity; strengthening of soil and fertilizer testing facilities to provide soil test based recommendations to farmers for improving soil fertility; ensuring quality control requirements of fertilizers , biofertilizers and organic fertilizers under Fertilizer Control Order, 1985; up-gradation of skill and knowledge of soil testing laboratory staff, extension staff and farmers through training and demonstrations; and promoting organic farming practices.

The components under Soil Health include trainings for fertilizer dealers, foreign nationals, fertilizer inspectors and fertilizer laboratory staff, setting up of new static Soil Testing Laboratories (STLs), setting up of new Mobile STLs, and strengthening of existing STLs, setting up of new Fertilizer Quality Control Laboratory (FQCL), strengthening of FQCL apart from trainings and demonstrations on balanced use of fertilizers. The components under organic farming include trainings for certification course on organic farming, refresher courses for analysts, field functionaries, trainers training, publishing of News Letters on organic farming and bio-fertilizers, setting up of fruit/vegetable/agro waste compost production units, setting up of bio-fertilizer production units, setting up of bio-fertilizer & Organic Fertilizer Quality Control Laboratories, promotion of organic inputs, training on organic farming, etc.

Under the scheme, setting up of 277 Soil Testing Laboratories (STLs), 3 Mobile STLs, strengthening of 38 STLs, strengthening of 8 FQCLs, 709 trainings, setting up of 1 Liquid Carrier based Biofertilizers productions unit, promotion of Micronutrients in 1000 ha and setting up 6308 Mini Soil Testing Labs have been approved during 2016-17 under SHM component.

A new scheme "Soil Health Card" has been approved for implementation during the remaining period of 12th Plan to provide 14 crore Soil Health Cards to the farmers in the country. Soil Health Card will provide information to farmers on soil nutrient status of their soil and recommendation on appropriate dosage of nutrients to be applied for improving soil health and its fertility. Soil health card will be issued every 2 years for all land holdings in the country so as to promote balanced and integrated use of plant nutrients. Under the scheme 239.73 lakh soil samples collected and 444.58 lakh soil Health Cards issued by States, as on 03.01.2017.

Pradhan Mantri Krishi Sinchayee Yojana (PMKSY)

Pradhan Mantri Krishi Sinchayee Yojana (PMKSY) is being implemented with the objective of developing a long term solution for mitigating the effect of drought and increasing area under irrigation with Motto of '*Har Khet Ko Pani'*. This programme is being implemented in mission mode by three ministries with Ministry of Water Resources, RD & GR leading the mission. With an outlay of Rs. 50,000 crore for a period of 5 years (2015-16 to 2019-20), the scheme has been conceived amalgamating ongoing schemes viz. Accelerated Irrigation Benefit Programme (AIBP) of the Ministry of Water Resources, River Development & Ganga Rejuvenation (MoWR, RD&GR), Integrated Watershed Management Programme (IWMP) of Department of Land Resources (DoLR) and the On Farm Water Management (OFWM) of Department of Agriculture, Cooperation and Farmers Welfare. The major objective of PMKSY is to achieve convergence of investments in irrigation at the field level, expand cultivable area under assured irrigation, and improve on-farm water use efficiency to reduce wastage of water, enhance the adoption of precision irrigation and other water saving technologies (More crops per drop), promote sustainable water conservation practices etc. The aim of PMKSY is not only the creation of assured irrigation but to create protected irrigation by using rainwater by "*Jal Sanchay*" and "*Jal Sinchan*". The website of PMKSY (http://www.pmksy.gov.in) with static features and Management Information System (MIS) with multi-nodal access features has been operationalized with integrating and capturing information of various Departments/ Ministries for PMKSY.

Paramparagat Krishi Vikas Yojana (PKVY)

Paramparagat Krishi Vikas Yojana (PKVY) is one of the schemes under National Mission for Sustainable Agriculture (NMSA) to promote certified organic cultivation in 2 lakh ha covering 10,000 clusters. Financial assistance of Rs 50,000 per ha per farmer is provided in 3 years.

Participatory Guarantee System (PGS) of certification is promoted in PKVY clusters. The financial assistance is provided to clusters on different sub components for mobilization of farmers, organic seeds, to harvest biological nitrogen etc. It includes different components such as:

(i) Mobilization of farmers: training of farmers and exposure visit by farmers.

(ii) Quality control: soil sample analysis, process documentation, inspection of fields of cluster members, residue analysis, certification charges and administrative expenses for certification.

(iii) Conversion practices: transition from current practices to organic farming, which includes procurement of organic inputs, organic seeds and traditional organic input production units and biological nitrogen harvest planting etc.

(iv) Integrated manure management: procurement of Liquid Bio fertilizer consortia/Bio pesticides, Neem cake, Phosphate Rich Organic Manure and Vermicompost.

(v) Custom hiring centre charges: to hire agricultural implements as per SMAM guidelines.

(vi) Labeling and Packaging assistance & transport assistance.

(vii) Marketing through organic fairs.

Agricultural Technology Management Agency (ATMA)

The Scheme 'Agricultural Technology Management Agency (ATMA)' implemented since 2005 has now been included as a component of the Sub-Mission on Agriculture Extension (SMAE) under National Mission on Agricultural Extension and Technology (NMAET) with some cost revisions and additional components. It is now under implementation in **652 districts of 29 states & 3 UTs** of the country. The scheme promotes decentralized farmer-driven and farmer accountable extension system through an institutional arrangement for technology dissemination in the form of an ATMA at district level.

Under the scheme grants in-aid is released to states with an objective to support State Governments efforts of revitalization of the extension system and making available the latest agricultural technologies in different thematic areas to increase agricultural production through extension activities viz. Farmers Training, Demonstrations, Exposure Visits, Kisan Mela, Mobilization of Farmers Groups and Setting up of Farm Schools. Through these activities, latest agriculture technologies are disseminated to farmers of the country.

In order to promote key reforms under the scheme, ATMA Cafeteria 2014 continues to support activities in line with the following policy parameters:

Multi-agency extension strategies: At least 10% of allocation on recurring activities at district level is to be used through non-governmental sector viz. NGOs, Farmers' Organization (FOs), Panchayati Raj Institutions (PRIs), para-extension workers, agri-preneurs, input suppliers, corporate sector, etc.

Farming system approach: The activities specified in the cafeteria are broad enough to promote extension delivery consistent with farming systems approach and extension needs emerging through Strategic Research and Extension Plan (SREP).

Farmer centric extension services: The cafeteria provides for group-based extension and it has necessary allocation for activities related to organizing and supporting farmer groups. In order to supplement these efforts, a provision for rewards and incentives to the best organized farmer groups has also been provided.

Convergence: The Strategic Research and Extension Plan (SREP) and State Extension Work Plans (SEWP) would also be mechanisms for ensuring convergence of all activities for extension. At present, resources for extension activities are being provided under different schemes of Centre/ State Governments. It is mandated that the SEWP, submitted by the State Governments for funding under the scheme shall explicitly specify the activities to be supported from within the resources of other ongoing schemes as well as from this scheme.

Mainstreaming gender concerns: It is mandated that at least 30% of resources on programmes and activities are utilized for women farmers and women extension functionaries.

The 'important features' of the 'Cafeteria of Activities' supported under Modified ATMA Scheme are as follows.

- Support for specialists and functionaries at State, District and Block Level
- Innovative support through a 'Farmer Friend' at Village Level @ 1 Farmer Friend per two villages
- Farmers Advisory Committees at State, District and Block levels Farm Schools in the field of outstanding farmers being promoted at Block/ Gram Panchayat level by integrating the Progressive farmers into Agricultural Extension System (AES) (3-5 Farm Schools/ block).
- Farmer-to-farmer extension support at the village level to be promoted through Farmers' Group.
- Funding window provided at both State and District levels for implementing innovative extension activities are not specifically covered under the Programme.
- Farmers' Awards instituted at Block, District and State levels.
- Community Radio Stations (CRS) to be set up by KVKs being promoted under the Programme.
- For Non-Governmental implementing agencies, States have been given the flexibility of having State Extension Work Plans prepared and approved at the State level. At least 10 per cent of outlay of the Programme is to be utilized through involvement of non-governmental sector. Non-governmental implementing agencies (excluding the corporate sector) are also eligible for service charge up to a maximum of 10 percent of the cost of the extension activities implemented through them. Apart from other NGOs financial assistance is also available for implementation of extension activities through agri-preneurs.

Kisan Call Centres (KCCs)

The KCC Scheme was launched on 21st January 2004 to provide answer to farmers' queries on agriculture and allied sectors through toll free telephone lines. A country wide common eleven digit number **'1800-180-1551'** has been allocated for KCC. The replies to the queries of the farming community are being given in 22 local languages. KCCs operate from 14 locations in the country covering all the Sates and UTs. Calls are attended from 6.00 am to 10.00 pm on all 7 days of a week.

Since inception of the scheme till 30th November, 2016 over 300.63 lakh calls have been registered in the KCCs. During the current year around 44 lakh calls have been received up to 30th November, 2016. In order to make farmers aware of this facility, audio and video spots on Kisan Call Centres have been broadcast/telecast through All India Radio, Doordarshan and private television channels. A Kisan Knowledge Management system (KKMS) has been created at the backend to capture details of the farmers calling KCCs. Modified call Escalation Matrix has also been put in place. If the queries are not answered by FTA is escalated to concerned Block Level Officer for replying the query through KKMS interface within given time frame.

The Kisan Call Centre scheme has been recently restructured and strengthened, with a good number of unique features viz. provision of IPPBX, 100% call recording; call barging; voice mail service; customized IVRS; call conferencing through the experts; playing state specific advisories during call wait time and SMS to caller farmers giving a gist of answers given by the KCC Agent now known as Farm Tele Advisor. The farmer calling KCC can also register for receiving SMSs from experts on the subject area and their providing and also for receiving regular updates on mandi price of selected mandis and crops.

National e-Governance Plan in Agriculture (NeGP-A)

The Government is implementing a state sector plan scheme (erstwhile centrally sponsored) Mission Mode Project (MMP) named **National e-Governance Plan in Agriculture (NeGP-A)** for helping farmers to access information related to latest technology. This project has been approved at a total cost of Rs. 858.79 crore for implementation of this scheme in entire country. Dissemination of information to the farmers has been aimed through various delivery channels including Common Service Centres, Web Portals, SMSs and Kisan Call Centres and Mobile apps in twelve identified clusters of services under the project are under implementation.

Services Rendered

The services include Information on Pesticides, Fertilizers and Seeds; Soil Health; Information on crops, farm machinery, training and Good Agricultural Practices (GAPs);

Weather advisories; Information on prices, arrivals, procurement points, and providing interaction platform; Electronic certification for exports and import; Information on marketing infrastructure; Monitoring implementation / evaluation of schemes and programmes; Information on fishery inputs; Information on irrigation infrastructure; Drought Relief and Management; Livestock Management.

mKisan portal

This Portal subsumes all mobile based initiatives in the field of agriculture & allied sectors. It brings together SMS (both Push and Pull), Interactive Voice Response System, Mobile Apps and Services. Officers, Scientists and Experts from all Organizations and Departments of the GoI and State Governments (including State Agriculture Universities (SAUs) , Krishi Vigyan

Kendras (KVKs) and Agro –Meteorological Field Units (AMFUs) are using this Portal all over the country and using this Portal for disseminating information(giving topical & seasonal advisories and providing services through SMSs to farmers in their local languages) on various agricultural activities to registered farmers, during FY 2015-16, more than **780 crore** SMSs and during FY 2016-17 more than 480 crore SMSs have been sent till 31.12.2016 to farmers by all agencies/organization/departments in agriculture and allied sectors down to Block level throughout the country since its inception July-2013.

The content may include information about the Schemes, Advisories from Experts, Market Prices, Weather Reports, and Soil Test Reports etc. The farmers can register for this service by calling Kisan Call Center on the toll free number 1800-180-1551 or through the Web Portal/SMS. As part of multiple channels of delivery of information to the farmers, DAC&FW has already launched five mobile apps (Kisan Suvidha, Pusa Krishi, Crop Insurance, Agrimarket & Bhuwan Hail storm). It has also been decided to make complete information and services available to the farmers through suitable Mobile Apps. Few important Mobile Apps developed by Department are:

(a) **Hail Storm application:** Farmer or other stakeholder may upload hail storm photograph with GPS tagging.

(b) **Crop Insurance**: Famer can know insurance premium, notified area on the mobile.

(c) **Agri Market**: Farmer can know the prices of various crops in the mandis near him.

(d) **Kisan Suvidha:** It has a simple interface and provides information on **five critical parameters** weather, input dealers, market price, plant protection and expert advisories.

An additional tab directly connects the farmer with the Kisan Call Centre where agriculture experts answer their queries. Unique features like extreme weather alerts and market prices of commodity in nearest area and the maximum price in State as well as India have been added to empower farmers in the best possible manner.

Kisan Suvidha mobile app is presently functioning in English & Hindi with a facility of language translation. States have been given a facility of translating the app in their own language. This translation facility will help in extending the outreach of this app and thus more and more farmers can be benefited throughout the length and breadth of the nation. It is available in Google Play Store and on mKisan – Mobile Apps section (mkisan.gov. in). Till December 2016 more than 3.7 Lakh users have downloaded the App.

(e) **Pusa Krishi-Mobile App:** Pusa Krishi Mobile App has been launched recently by the Government. The objective of this app is to help farmers to get information about technologies developed by IARI, which will eventually help in increasing returns to farmers.

(f) **CCE App:** for undertaking crop cutting experiment.

Strengthening of IT Apparatus in Agriculture and Cooperation in the States and Union Territories (AGRISNET)

The Government is implementing a Central Sector Plan Scheme "Strengthening/ promoting agricultural Informatics & Communications", of which one of the components is AGRISNET. The objective of AGRISNET is to provide improved services to the farming community through use of ICT. Under the AGRISNET Scheme funds are released to the State Government concerned for provisioning of software and hardware systems including networking, data digitization, and manpower training to ensure computerization up to Block level 26 States have so far availed funds released by the Government under AGRISNET.

National Crop Insurance Programme (NCIP)

Keeping in view the risks involved in agriculture and to insure the farming community against various risks, Ministry of agriculture introduced a crop insurance scheme in 1985 and thereafter brought improvements in the erstwhile scheme(s) from time to time based on the experience gained and views of the stakeholders, States, farming community etc.

To enlarge the coverage in terms of farmers, crops and risks, National Agricultural Insurance Scheme (NAIS) was notified/ implemented with effect from 1999. To make the crop insurance schemes more farmers' friendly, a re-structured Central Sector crop insurance scheme namely, "National Crop Insurance Programme (NCIP)" was implemented by merging erstwhile Pilot schemes of Modified National Agricultural Insurance Scheme (MNAIS), Weather Based Crop Insurance Scheme (WBCIS) and Coconut Palm Insurance Scheme (CPIS) (as its components) with some improvements for its full-fledged implementation from Rabi 2013-14 season throughout the country.

National Agricultural Insurance Scheme (NAIS) was to be discontinued after implementation of NCIP from Rabi season of 2013-14. However, on the representations and at the option of States, NAIS was also allowed for implementation up to 2015-16. The erstwhile crop insurance schemes have further been reviewed in consultation with various stakeholders including States/UTs. and new scheme namely, **Pradhan Mantri Fasal Bima Yojana (PMFBY)** has been approved for implementation from Kharif 2016 along with pilot Unified Package Insurance Scheme (UPIS) and Restructured Weather Based Crop Insurance Scheme (RWBCIS).

The erstwhile crop insurance schemes have recently been reviewed in consultation with various stakeholders including States/UTs. As a result of the review, a new scheme **"Pradhan Mantri Fasal Bima Yojana (PMFBY)** has been approved for implementation from Kharif 2016 along with pilot Unified Package Insurance Scheme (UPIS) and restructured Weather Based Crop Insurance Scheme (WBCIS). Under the PMFBY, a uniform maximum premium of only 2% will be paid by farmers for all *Kharif* crops and 1.5% for all Rabi crops. In case of annual commercial and horticultural crops, the maximum premium to be paid by farmers will be only 5%. The premium rates to be paid by farmers are very low and balance premium will be paid by the Government, to be shared equally by State & Central Government, to provide full insured amount to the farmers against crop loss on account of natural calamities. There is no upper limit on Government subsidy.

National Horticulture Mission (NHM)

This Centrally Sponsored Scheme was launched in the year 2005-06 aims at the holistic development of horticulture sector by ensuring forward and backward linkage through a cluster approach with the active participation of all stake holders. 384 districts in 19 States and 4 Union Territories were covered under NHM. Eighteen (18) National Level Agencies (NLAs) have also been included for providing support for developmental efforts which require inputs at the National level. Supply of quality planting material through establishment of nurseries and tissue culture units production and productivity improvement programmes through area expansion and rejuvenation, technology promotion, technology dissemination, human resource development, creation of infrastructure for post harvest management and marketing in consonance with the comparative advantages of each State/region and their diverse agro-climatic conditions were major programmes implemented during 2014-15. During 2016-17, as on 31st December, 2016, funds to the tune of Rs. 605.61 crore have been released to States/Union Territories and National Level Agencies implementing NHM scheme against BE of Rs.890.00 crore.

Horticulture Mission for North East and Himalayan States (HMNEH)

Ministry of Agriculture & Farmers Welfare, Government of India has been implementing a Centrally Sponsored Scheme - Horticulture Mission for North East and Himalayan States (HMNEH) earlier known as “Technology Mission for Integrated Development of Horticulture in North Eastern States since 2001-02. During the X Plan (2003-04), the scheme was further extended to three Himalayan States namely: Himachal Pradesh, Jammu and Kashmir and Uttrakhand. The Mission covers entire spectrum of horticulture, right from planting to consumption, with backward and forward linkages. With effect from 2014-15, HMNEH scheme has been subsumed under the Mission for Integrated Development of Horticulture (MIDH).

During 2015-16 (As on 18th Dec, 2016), an area of 3823 ha has been covered under various horticultural crops besides 934 ha of area under old and senile orchards has been rejuvenated. An area of 3057 ha has been brought under organic farming. Under protected cultivation, an area of 58.22 ha has been covered besides 16446 ha under Integrated Pest Management, 5 Nurseries have been established and 235 (nos.) under harvesting structures have been created. 46 PHM infrastructures and 21 rural markets have been set up.

During 2016-17, funds to the tune of Rs. 181.29 crore have been released to States implementing HMNEH scheme against BE of Rs.320.00 crore.

National Agro-forestry and Bamboo Mission (NABM)

National Agroforestry & Bamboo Mission (NABM) scheme is being implemented in all States except Haryana in the country since 2006-07. The Mission envisages promoting holistic growth of bamboo sector by adopting area- based, regionally differentiated strategy and to increase the area under bamboo cultivation and marketing. Under the Mission, steps have been taken to increase the availability of quality planting material by supporting the setting up of new nurseries, strengthening of existing tissue culture units. Besides research projects have also been funded for identification of superior bamboo clumps for various bamboo species and for establishing nursery in their respective agro-ecology to have best planting stock. To address forward integration, the Mission is taking steps to strengthen marketing of bamboo products, especially those of handicrafts items.

Coconut Development Board (CDB)

The Coconut Development Board was established by an Act of Parliament in 1979 and came into existence on 12.01.1981. The Board is implementing various programmes for the development of coconut sector which includes action plan like production and distribution of planting material, expansion of area under coconut and integrated farming in coconut holding for productivity improvement and a technology mission on coconut.

National Horticulture Board (NHB)

The National Horticulture Board was established in 1984 as an autonomous registered society under the Department, with its headquarters in Gurgaon and 35 field offices located all over the country. Its main objectives are the creation of production hubs for commercial horticulture development, post-harvest infrastructure and cold chain facilities, promotion of new technologies, introduction and promotion of new crops, and promotion of growers' associations. The significant feature is that there has been improvement of productivity of horticulture crops. Area, production and productivity of horticulture crops during past 9 years are given in the following table.

Table 6: Areas, Production and Productivity of Horticulture Crops

(Area : m Ha, Production: m. MT, Productivity: MT/ha)

Year	Area	Production	Productivity
2007-08	20.2	211.0	10.4
2008-09	20.5	214.4	10.5
2009-10	20.8	223.2	10.7
2010-11	21.8	240.4	11.0
2011-12	23.2	257.3	11.1
2012-13	23.7	268.8	11. 3
2013-14	24.2	277.4	11.5
2014-15	23.4	280.99	12.01
2016	23.7	283.36	11.91

The Board is implementing the following schemes:

- Development of Commercial Horticulture through Production and Post-Harvest Management
- Capital Investment Subsidy for Construction/ Expansion/ Modernization of Cold Storages and Storages for Horticulture Produce
- Technology Development and Transfer for Horticulture Produce
- Market Information Service for Horticulture Crops

- Horticulture Promotion Services/ Expert Services and Strengthening Capability of NHB.

The area under fruit crops during 2014-15 was 6.2 m. ha with a total production of 86.3 million MT. India has retained its status as the second largest producer of fruits in the world. The country is first in the production of fruits like mango, banana, sapota, pomegranate and anola.

Vegetables are an important crop in horticulture sector, occupying an area of 9.5 million ha during 2014-15 with a total production of 167.1 million tons with average productivity of 17.6 tons/ha. In fact vegetables constitute about 60% of horticulture production. During the period (2007-08 to 2014-15), area and production of vegetables increased by 20% and 30% respectively. India continued to be second largest producer of vegetables after China. Our country is a leader in production of vegetables like peas and okra. Besides, India occupies the second position in production of brinjal, cabbage, cauliflower and onion and third in potato and tomato in the world. Vegetables such as potato, tomato, okra and cucurbits are produced abundantly in the country.

Interventions in horticulture in the country, have led to increase per capita availability of fruits from 158 gm/person/day in 2007-08 to 189 g/person/day in 2014. Similarly, per capita availability of vegetables has increased from 309 g/person/day in 2007-08 to 357 g/person/day in 2014.

India has also made noticeable advancements in production of flowers, particularly cut flowers, which have a high potential for exports. Floriculture during 2014-15 covered an area of 0.5 million ha with a production of 1.66 m. MT of loose flowers and 2.14 m. MT of cut flowers. There has been phenomenal growth in cut flower production, which is finding place in export market.

India is the largest producer, consumer and exporter of spices and spice products, the total production of spices during 2014-15 was 5.7 m. MT from an area of 3.19 m. MT. India continues to be the largest producer and exporter of cashew nut in the world. As per the final estimates 2014-15, the production of cashew nut in India during 2014-15 is 7.25 lakh MT from an area of 10.27 lakh hectares unit area productivity of 706 kg/ha. The cashew processing industry requires 17 lakh MT per annum of raw cashew nut. There is a wide gap between our production and demand of raw cashew nut and India imports the rest of its requirements from other countries.

In India, cocoa is cultivated in four southern states of Kerala, Andhra Pradesh, Tamil Nadu and Karnataka with little plantation status mainly as an intercrop in coconut and areca nut gardens. It is also cultivated in shady conditions of rubber plantations in few localities of Kerala. Cultivation of cocoa is gaining momentum and as per the final estimate 2014-15, India is presently produced

16050 MT of cocoa from an area of 78000 ha. The country is importing about 40 % of the demand for cocoa and the demand is increasing in the country by 15 % annually. India imports rest of its requirement from other cocoa producing countries. There is lot of scope and demand to further increase in the area and production of cashew nut and cocoa. The Directorate of Cashew nut and Cocoa Development is implementing development programmes under MIDH to increase the area and production of these two horticultural crops in the country.

Conclusion

The agricultural sector is of basic accent and index of development for the region. It is knowledge of alteration to a market economy, with abundant changes in the social, legal, structural, advantageous and accumulation set-ups, as is the case with all additional sectors of the economy. These continuous changes accept that has been accompanied by in decline in agricultural production for which a lot of changes are taken place in the countries, and accept afflicted as well the national accumulation sectors of the region. The research and development are therefore to be updated occasionally to get information on the progress. Some favourable types of climates and additional adverse factors, including an actual advanced academic accumulation due to the almost low demographic burden projected for the future, should be accessible to affected problems of food crisis in the arena as a whole, and even to use this arena to accommodate additional food to food-deficient regions. Opportunities have to accordingly be created to knowledge towards these results. In the system to know the capital constraints affecting the development of the national and state level food problem that are mentioned here, efforts by all national and all local stakeholders and institutions should come together in accumulation of bio genetic knowledge and its management.

Some pertinent short questions

1. What is full form of HMNEH?
2. What are GVA and GDP?
3. What do you know about capital formation in agriculture and allied sector?
4. Write function of NHB?
5. What do you know about mKisan portal?

17

Indian Agricultural Concerns and Future Prospects, Agricultural Vision, The Future and Challenges Ahead

Introduction

Today, concerns regarding environmental safety and sustainability of land productivity are increasing amongst the scientists, administrators and environmentalists. It is doubted whether the strategy adopted during the so-called green revolution era could be continued any longer under the challenging conditions of this new century. Already, a section of people in the world is questioning the propriety of conventional agriculture, and a few of them are advocating alternative practices that are perceived to lay foundation for sustained production. On these lines, systems like alternative agriculture, natural farming and organic farming were proposed at various conventions. However, the scientists harping on the success of green revolution continue to doubt whether such a system can really be functional, productive and meet the growing demands for agricultural products in this e-age. These emphasize the need to develop new strategy of living with the nature and nurturing it for sustainable production

Speculation on Indian agriculture

The spectacular observations made in the Economic Survey of 2015-16 that "Indian agriculture, is in a way, a victim of its own accomplished success and in abnormality the green revolution", shows the absoluteness of the agricultural planning at present and the calamity that has been wreaked by the green revolution. The green revolution, which is generally characterized by the addition of high-yielding varieties of seeds and fertilizers uses, assuredly added the higher quantity of crop coverage considerably. There has been acceptance of destroying ecological balance furnishings in the composition in felling down water table, discharge of greenhouse gases, and the contamination of ground water. Needless to say, the agricultural area is in an accompaniment of distress, which is acutely affecting peasants and bordering farmers, and burning action interventions are adapted to assure their interests. The government has responded to constitute basic panel, which will acclaim organization to increase the assets of farmers by 2022. This might be a frankly insistent target, if not wish to make addition of stagnated agronomical advance in open face to be made. However, there are flourishing hurdles that to be accepted if to wish accomplishment of this objective.

Rainbow revolution

The strategies is to achieve this objective of increasing production and productivity, it refers to a sustainable agricultural revolution where equal focus is to be given on minimizing cost of production while increasing produce with less environmental adverse impacts, and lesser use of chemical fertilizers and maintaining soil health.

The native above obstructions are supposed to affect is collapsed or downtrend in productivity that reveals from the report where India's documents of crop yield per hectare is far beneath than that of affluent countries (including several low assets countries), but the deviation is huge if it compared to China. For instance, our evident shows crop yield per hectare is 39% on an average below than that of China and for rice this volume is 46%. Even Bangladesh, Vietnam and Indonesia volume are bigger than India in case of rice yield. Further, there is a huge inter-regional variation where the wheat and rice crop yield from Haryana and Punjab is abundant greater than from the added states.

In adjustment to argumentative this breakdown profusion of blockade there is a need to charge to adviser to make change over by trustworthy directions to get more production from wheat-rice. Since wheat and rice accompanying with added crops are backed by minimum support prices (MSP) and ascribed subsidy (water, fertilizer or power), there is a huge scope for the farmers in the dry regions of North-West India to be plentiful production of these crops. These crops are not alone ascribing intensive, but as well accepted to restore ecology after-effects in the checking water table in low lying areas and resist discharge of green gases. However, this is not adequate and has to be complemented with huge investment in accessible infrastructure. For example, due to the rice milling industry in Haryana and Punjab, there is now an able accustomed market for different quality varieties of rice that as well incentivizes farmers to breed paddy seed. Until such an exchange is not created for added cereals and pulses, farmers are ridiculous to accomplish the changes in cereals and pulses production.

Per drop more crop and income

The addition to above barrier stated above, there are absences of two assets for agricultural change over these are (1) cultivable land and (2) water. Per-capita cultivable land per is decreasing because of the breaking of farms size due to ever-increasing population, India as well has availability of per capita water is also decreasing trend as compared to other agrarian countries. This painful situation arises and affronted because of non-judicious water application in growing crops, which is apparent by its affection of non-replenishment.

Given this scenario, it is time to accomplish an introduction to micro irrigation system so that the able and authentic use of deficient water assets can be made. A survey conducted by the National Mission for Sustainable Agricultural on Micro Irrigation in 64 districts of 13 states (Andhra Pradesh, Bihar, Chhattisgarh, Gujarat, Haryana, Karnataka, Maharashtra, Odisha, Rajasthan, Tamil Nadu, Sikkim, Uttar Pradesh and Uttarakhand), reveals that there were convincing not only reductions in the use of water and fertilizer but the crop of crops added up to 45% in wheat, 20% in gram and 40% in soybean. However, relatively higher costs turn away farmers to accept this technology. The big farmers can easily afford the expenditure and accepted these water saving technologies wherein the government should come forward giving subsidies to small farmers for readily acceptance of this technology.

Further, it was also noted that, due to independent installation charges, the prices of electricity and/or fuel requirement are higher. Hence, to make the programme on more crop and income per drop, introduction of micro-irrigation is inevitable and wherein water charging at minimum costs affordable by the small famers should be looked into as the technology is accessible.

Opening up of the markets

National Agronomical Action of 2000 declared that concealed area that accorded acknowledgment through arrangement agricultural inputs and land coverage leasing arrange to agree accelerated location-specific and proven technology transfer for crop production, basic arrival and assured markets for sale the agricultural produce. However, there has not been any convincing accord to solve the issues by the concealed area in agriculture.

One of the above factors that have stricken secret issue from entering the agronomical area is the continued awaiting for broad markets, which are adapted by the Agricultural Production Management Committee (APMC) Act. The AMPC Act is the guard the farmers to advertise their products in government-controlled business domain.

In 2003, however, the Union Government mooted a model APMC, but as acclaimed by the assignment force on agricultural constituted by NITI Aayog, this has not been implemented by affluent states in east India. Therefore, to access larger area accord in agriculture, it is sensitive to eliminate these entry barriers. Further, although the government has launched the National

Agricultural Market, which provides a cyber-banking coverage to advertise their products and facilitate marketing of produce anywhere in India, it is yet to be apparent whether farmers can in fact acquire allowances to access any good from this platform.

Future Research and Development

In the context of above barriers to advocacy of land coverage in greater quantity, it has become the synopsis of new technologies and breakthrough in agricultural scenario. National Agricultural Commission played an important role in the green revolution, in contempt years where there hasn't been any above advance in agricultural research. One of the capital documents for this is the involvement of banking resources. If analyze the summary of the allotment of agronomical GDP spending on examining development in Asia, again these would be revealing. India spent 31% of its GDP on agriculture and development in 2010, in the aforementioned year China spent about double than this amount. Even our neighbour Bangladesh spent 38% of its GDP on agriculture and development in that year. As an aftereffect of this capability crisis there has not been circulation of new agronomical innovations and practices that is analytical for adequate land productivity.

Further, there is a synopsis of incorporation of acceptance in advancing access in agriculture. Economic Survey noted, even in respective states agricultural is almost added important (as severe by their allotment of agricultural GDP), agricultural apprenticeship is abnormally weak if clear-headed by the volume of acceptance enrolled in agricultural universities. There has as well not been any above addition from the larger area appear in agriculture and development. Government should appropriately play an important role by giving them incentives to play role in agricultural growth and development. Many accepted casting doubts over the desire stand of the government to double the income of the farmers by 2022. Administrator of Commission for Agricultural Costs and Prices noted, acceleration of total incomes of farmers would be a "miracle of miracles", as it would be taken an admixture advance increase of 12% per annum. Further, India spend is as well disbeliever of government desire as their growth in agriculture shows that afterwards adjusting for rise in costs, an Indian farmer's income rose alone 5% per year over a decade (2003-2013). All this, in these ways, it paints a severe account of approaching of Indian agriculture. If we about wish to save the approaching of our farmers and assuredly cure the ills of Indian agriculture, above action interventions to be accepted redressed at the earliest. The arrangement resembles the agricultural acclimated to be plentiful vegetables production in field and in greenhouses. Every half-hour an anxiously calibrated thrashing of water based on the calculations, and alloyed with an adapted dosage of fertilizer if scheduled, is pushed through the tapes, carrying an absolute admixture. The pulses alternating amid one ancillary of the timberline blocks and the other, which associate has been apparent encourages water uptake. Before this arrangement was in place, it would be accepted dry/rainfed coverage about already a weak. With the new little-but-often technique, uses 20% less water than acclimated to. Understanding a crop's arrangement as well agency that extraction itself can be fabricated added precise.

Such complex changes, in hardware, software and "livewire", are expensive and extensive in the agricultural field and orchard. Fish farming will be acted as well get an addition from them. And composed and planned horticulture, already the lot of controlled and complete display of agriculture, is about to become yet added so.

In the cut short run, these improvements will be acted as addition farmers' profits, by sharp costs and buildup yields, and should as well be accounted consumers (meaning anybody who eats food) in the composition of lower prices. In the best run, though, they may advice to accommodate the acknowledgment to a more burning question: how can the million are fed in approaching after putting irreparable throbbing on the Earth's soils and oceans? Amid now and 2050 the planet's population is acceptable to acceleration to 9.7 billion, from 7.3 billion now. Those humans will not alone charge to eat, they will wish to eat bigger than humans do now, because by again a lot of are acceptable to accept middling incomes, and affluent will be vigorous off on consumption.

Gap existing that required that specific intervention(s)

1. Reversing sick soil to healthy soil through low cost approaches in existing indigence;
2. These gaps exist for other livestock also with introduction of exotic breeds of goat suitable for stall feeding;
3. Standardization of methodologies/approaches for generating gainful rural self-employment through improved agricultural technological interventions;
4. Low cost organic farming practices;
5. Introduction of crop diversification, change in poor breeds of animals/ birds, introduce farm mechanization, replace mono-cropping, replacement quality with proper rate; and judicious water use avoiding stress and good marketing infrastructure facilities are some important areas;

The major components of the project taken up

(a) Intensification and diversification of agricultural activity through integrated farming system approach comprising varietals improvement, seed replacement, farm-based enterprises for self-employment;

(b) Livestock improvement, increasing per unit profitability and integration;

(c) Soil health improvement by promoting organic production system, IPM, mass composting, encouraging use of homemade bio-pesticides, bio-fertilizers; and

(d) Human resource development for sustenance of the proposed activities;

Participatory forest management: In a developing country like India, development encompasses several wide ranging issues. The prime aim of development for a benefit country is peoples welfare and for public good. In fact every country throughout the world strive towards this objectives and so also all the countries in the region. Forest is a part of the environment. Development and environment both are inseparable. Man lives in a given environment and strives to improve his lot by certain actions within it in the name of development. Every act of man affects environment. People's well being is the ultimate goal of all governments and all development policies and programmes are usually tailored to cater to that objective. Unlike everything is changing land remains and will continue to remain limited. Growth demands space and forest is the only area having some geographic spread. This perhaps has been also the cause of liberal use of its area and extent. The increasing need and influence of other sectors has made forestry sector a diminishing entity. In the economic terms and financial accounting system externalities (social costs) are not considered very often. Services from forest such as soil protection and productivity, regeneration of water resources and improvement of hydrological cycle, all the forest products other than timber and bamboo, carbon sink and bio-diversity have immense value of far reaching consequences. Such benefits flowing from forests are not only having local but regional and global importance. Large scale industrialization does not mean development if it does not pay due attention to environment. Further development has to have social content in every sphere. The material progress has to be blended with old values, ethics and morals. Earlier no one talked of environment in development projects and activities. The craze for development subdued the simmering after effects or fallout of those activities which do not seem visible initially in the beginning. Forests destroyed to make way for development and industries which later polluted air and water causing several difficulties and problems to the people in general. In this regard the examples of devastating effect of indiscriminate development are not wanting. Government has taken care against such reckless development and ruthless predatory exploitation of forest (natural resources) and environment. Otherwise the cure becomes much more cost prohibitive than the disease itself. Realization has come for the old saying "Prevention is better than cure".

Visualizing the scenario for the future as well as for the incoming century, Government of India with its age-old wisdom is the first country to make constitutional provision for protection and improvement of environment in 1976. Forestry and wildlife are in the concurrent list of the Constitution of India. Ministry of Environment and Forests is the nodal administrative structure at central level for planning, promotion and co-ordination. Ministry is responsible for policy formulation, monitoring, and co-ordination with Government and International agencies. Environmental impact assessment of developmental works, prevention and control pollution, research education and training and providing financial and technical assistance to states, NGO's and others through various schemes are also the responsibility of the Ministry. Conservation and development efforts have been facilitated by several legislative and administrative instruments. Government of India through such measures has tried to ensure "Development without Destruction". The 1988 'National Forest Policy' is a landmark instrument with wide ranging coverage and focusing on conservation efforts in Indian forestry coupled with sustainable utilization and people's participation. The National Conservation Strategy was adopted in 1992 laying down stress on environmental considerations. The principal aim of National Forest Policy (1988) is to ensure environmental stability and maintenance of ecological balance including atmospheric equilibrium which is vital for sustenance of all life forms like human, animal and plant. The derivation of direct economic benefit must be subordinate to this principal aim. The policy lays down the overriding considerations for the environmental concerns rather than commercial interests. Accordingly, the conservation of bio-diversity, genetic resources, species, their habitats and production of wood for local consumption and use as well as for meeting the industrial requirements on long term basis are important factors determining sustainability. India has unique and diverse forest ecosystems, some being fragile and critical needing special treatment. Most of the India's forests being under the control of government are managed on scientific lines with detailed field work, inventory and mapping with due regard to principles of sustainable development. The policy document stipulates that holders of customary rights and concessions in forest areas should be motivated and encouraged to identify themselves with protection and development of forest from which they have been deriving benefits. Circular dated 1st June 1990, reflects government's commitment to manage the forests through participatory process which envisages active involvement of village communities, specially women and weaker sections of society and voluntary agencies. Rights and concession of the tribal and other poor living within and near forests are protected and their *bonafide* domestic requirements are regarded as first charge on forest produce. In spite of general denudation of forests and lack of regeneration there has been at places instances of good regeneration in natural forest by changing forest management practices involving more effective collaboration between government forest staff and rural communities having willingly controlled their access to natural forest for grazing and cutting. This is an emergence of new approach hitherto not given adequate attention and importance. It is an integrated approach for ecological security and stability.

Adoption of participatory forest management with communities providing them incentives and support (both technical and financial) has wide acceptance now and is also more cost effective. This philosophy of forest development is a silent social revolution and has considerably reduced general degradation of the commons. People have responded to the cry of forest "first care and then share" and thereby people themselves have developed a stake in forest protection. This process has generated a sense of belonging amongst people. They have voluntarily turned "from free user to regulated user of resources and from consumer to producer". Consequently forest management is done with a human face and for the people by harmonizing various parameters of forest management. In a way it is liberalizing and feminizing forestry. This has helped the Forest Department to feel oneness with society with a concept of social justice and distribution of rights. Joint forest management has become the forestry in the country is in transition from a statist to a more people centred approach. The country's forest policy as now today is talking in terms of empowering and involving people in managing forest locally. That participatory approach is a viable alternative to the traditional methods of forest management became known when the in the state of West Bengal tried this concept for the first time in the early 1970's. It started as an attempt by a harassed forest department unable to prevent the destruction of forests in spite of strenuous efforts at policing, to involve people in policing forests in return for a share in the harvest. In general people were depending on forest and degradation of forest continues with way flow of produce. People were only concerned with their rights and concessions without bothering for their responsibilities. In the process poverty became endemic to degraded forest areas. People were persuaded first to care and then manage the forest and share the benefit in shape of goods though the services and functions from the forest so protected was uniform for the entire community. Involving people in this collective endeavor has in cost-effective and very radical in approach.

Protection alone enhances the bio-diversity of the area and thereafter helps in equitable sharing of biotic wealth. The novel attempt had salutary and significant effect on forest and local society as well and made them more aware and concerned for the benefits of forests and optimal use of resources not only for own generation but for generations to follow. Rules of self-abnegation were framed by the local communities leading to the laying of the foundation for collective efforts at regenerating degraded forests. Such efforts both of local governments and village communities have been also taking place in other parts of the country. In one instance a villager narrated to the author about their case as "necessity is the mother of inventions" resulting in extensive degraded forest area regenerating by their voluntary protection. People realized the ecological advantages by their action contrary to legacy of ecological exploitation and generated ecological movement which contributed more for survival of poor. The causative factors of forest degradation were taken care of by use of this renewable resource with restraint care and caution. The success that these efforts met with led to formulate a strategy to involve people in collaborating with the forest department to regenerate forests and get a share of the benefits (harvest) from the forests so managed.

Joint forest management: In India participating in forest management is called Joint Forest Management (JFM). JFM involves the collaboration of the forest department, the people of the area and a local non-governmental organization (NGO) to serve as an interface between people and government. From this modest beginning, JFM has acquired the shape of a movement with over 15,000 village forests committees (VFCs) managing in collaboration with forest department's over 2.0 M ha. of degraded forests. The experience in the last few years has shown that many complex issues of social, political, administrative, legal and silvicultural are emerging and require to be addressed. This mechanism for enhancing forest area both within and outside legal limits of forest has a great future because of recognition of its utility and universal acceptance. The trust of the people has to be sustained to continue the process towards prosperity. It can be emphatically foretold that the coverage under JFM and involvement of people will increase several tunes of what it is now since the awareness for forest conservation is gaining ground at the base in rural India nearer the forest. The return from forest varies widely in the country depending on the locality and species occurring therein.

The practice of forest working from the beginning has a scientific base in which both present and future needs were considered without sacrificing the biodiversity existing in the local forest types. The imbalance crept in when the demand surpassed the supply impinging and impairing the existence of the very forest. Forest alternatives and substitutes have been found and more and more research is on to save the overuse of wood and wood products. Both the aspects of supply and demand have to he tackled to conserve the natural forest in its perfect glory. Conservation encompasses both protection and management. Legal and Administrative framework have to be strengthened with proper human resource development and with a paradigm shift in their approach to various problems of forest management. Modern tools like GIS (Geographical Information System) and MIS (Management Information System) have to be used to hasten the process of building up of lost ground in forest management. The forest management has to be "technically sound, environmentally compatible, commercially viable and socially acceptable" and its ultimate aim and approach has to be humane and people oriented. That would include both short term and long term gains.

Natural forest extent may not remain static in view of growing population and cattle pressure and its consequential welfare needs and demands which include several developmental projects like mining. Shifting cultivation, though likely to reduce, will have its toll on the forest land. The reality of the projection ultimately means less of forest and more per capita demand. That being the case efforts has to be made for enhancing forest productivity with adequate research supports. The natural regeneration in the forest can be supplemented with artificial regeneration wherever necessary to maintain the forest stocking to the optimal state. Timber harvesting from forests is linked to rural employment and economics. However, the importance of remote and inaccessible forests tracts as a source of timber will diminish considerably. Plantations will gain importance as a source of all forested raw materials in the 21st century. The NGO activists have contributed in no small measure to this change. But there will still be some tracts of natural forests from where it will be necessary to manage forest scientifically and to extract timbers of selected species or dimensions. The returns from logging of such areas would be high and would cost sufficiently low to make harvesting of natural forests competitive with those of the plantation industry. If timber has to increasingly come from plantations there would be a need to reappraise policies regarding the proportion of natural forests to plantation. An important change that has been seen in the developed western countries is the realization by activist NGOs that environmental amenity value of forests can be maintained even when timber harvesting is continued in old growth forests. Similarly industries and forestry professionals are recognizing the need to modify harvesting practices so as to reduce its adverse impacts. The extent and quality of the forest cover a country can maintain is linked to the biotic pressures it is subjected to. It is expected that forest cover in most developing economies will continue to diminish till they are able to bring down the rates of population growth to manageable levels, develop their economies and create avenues for alternative sources of employment for people who now derive their livelihood from forests and forestry when this happens the forest cover will cease to decline further. But the threshold levels at which the U-turn will take place will be different for different countries. Even in the country different states will have different area coverage in forest because of their geographical location, population and development status.

Assuming an end of deforestation or forest degradation and continuation of afforestation/reforestation/agro-forestry activities at the rate of one million hectares per year, forest resources, by 2010 AD, could be somewhat like this:

- Area of natural forests	50 million Ha.
-Protected areas	15 million ha.
-Area with production potential	30 million ha.
-Area of forest plantation	40 million ha.
-Area under agro-forestry/homestead forests	8 to 10 million ha.

If a sustainable development scenario with effective protection of existing forests and increased investment in creating new-manmade forests is followed, India can avert the developing crisis and take a growth path in forestry. Even if the productivity of existing plantations (which is about 4 cu.m./ha/yr.) is increased to a level of 10 cu.m./ha/yr. (i.e. the lower limit to be qualified as fast growing), it should be able to produce about 500 million cu.m./yr from plantations alone.

Implications of Projected Outlook

The supply/demand situation and resulting changes have several implications affecting investment, skill needs, employment, income, government revenue, product substitution and diversification, product distribution, new entrepreneurial activities, value addition, operational efficiency, community welfare, environmental conservation, trade, prices and more importantly institutions and institutional arrangements.

Afforestation/reforestation: It is simply clear from the foregoing discussions that India's forests are being exploited in excess of their sustainable productive capacity. This is in spite of the fact that the country has increased the pace of its reforestation programme in recent years. Between 1951 and 1997 the country has reforested 21.78 Mha of degraded forest and other public lands at a cumulative cost of Rs. 68,842.8 million (approx US $ 1.9 billion) as per details in Table 1.

Table 1: Progress of reforestation in the country

Plan period	Area reforested (M ha)	Cumulative (M ha)	Expenditure incurred (Rs. million)	Cumulative (Rs. million)
First plan (1951-1956)	0.05	0.05	12.8	12.8
Second plan (1956-1961)	0.31	0.36	68.6	81.4
Third plan (1961-1966)	0.58	0.94	211.3	292.7
Annual Plans (1966-1969)	0.45	1.39	230.2	522.9
Fourth plan (1969-1974)	0.71	2.11	443.4	966.3
Fifth plan (1974-1979)	1.22	3.33	1,072.8	2,039.1
Annual Plans (1979-1980)	0.22	2.55	371.0	2,410.1
Sixth plan (1980-1985)	4.65	7,20	9,260.1	11,670.2
Seventh Plan (1985-1990)	8.86	16.06	25,868.4	35,128.6
Annual plan (1990-1991)	0.75	16.81	5,999.5	41,128.1
Annual plan (1991-1992)	1.15	17.96	7,701.3	41,898.4
Eighth plan (1992-1997)	3.82	21.78	26,944.4	68,842.2

Source: MoEF

Biggest challenge of Indian agriculture is the Fluctuating Market. The market today will not be there tomorrow. 75-80% of the farmers suffer heavily with this problem.

Main reasons for it are: *lack of cold storage to store the produce for sufficiently long time; *lack of quick transportation facilities like good roads to be there in market at right time; and *similarly farmers are incapable to purchase good quality packaging materials.

Greatest irony here is that, so called minimum support price given by the Government is not satisfactory.

Conclusion

The challenges before agricultural advancement attempt of ecology, the concept of relationships within microbes and their environment; refers to agricultural assembly which will be maintained while not harming the environment; not alone will it affect several structure and amusing issues, about it: offers forward-thinking and economically applicable opportunities for (i) growers (ii) laborers (iii) shoppers and policymakers; Cultivation of crops using cover crops, organic manures, organic manures, crop rotations to rebuild the soil; Agriculture requires use of biological management, crop rotations, to administer weeds, insects and illness emphasizing on assortment of agricultural planning and management; Agriculture needs use of motion agricultural and pastures for location specific operations and assorted green crops affliction of animal well-being and to reduce the strange and off coverage inputs and abolishment of bogus pesticides and fertilizers and which altered actual good hormones and antibiotics; and focusing on renewable resources, clay and conservation, and management practices that restore advance and enhance ecological balance.

Some pertinent short questions

1. What is rainbow revolution?
2. What is crop yield and income per drop?
3. What are the major components of project proposal?
4. Write few words about research and development in future;
5. What do you know by 'prevention is better than cure'?

18

International Agricultural Research Setup

Introduction

International agricultural setup has appeared as it now taking all out consideration and agricultural heritage, crop voyage, crop classification and its significances. International cooperation has contended a big role in developing and strengthening the research system in Asian country. several developed countries like USA, UK, USSR, Canada, Australia, Japan, many European countries; Charitable establishments, etc.; altruist and Ford Foundations; varied International Agencies like FAO, UNDP, UNESCO, World Bank, etc.; and also the International Agricultural Research Centres below the consultatory cluster on International Agricultural Research (CGIAR) System have contributed extensively to the explanation for agricultural RESEARCH in Asian country. Spectacular achievements in increasing the food production have raised the country's image significantly, and also the bilateral arrangements have modified from the erstwhile donor-donee standing to relationship of equal partnership in research. The reciprocity and mutuality of interests with the less developed and in addition because the technologically advanced countries square measure the essence of international co-operation. The government has approved ICAR, motor-assisted by the board of directors of Agricultural Research and Education (DARE), to enter into bilateral cooperative agreements with many countries and agencies. The mode of collaboration ordinarily follows the pattern of: (a) exchange of germ-plasm of plant and animal origin; (b) exchange of scientific and technical information; (c) visits of scientists and experts; (d) coaching of scientists; and (e) infrastructure development.

Major avenues of international collaboration

International collaboration has provided a mechanism to draw upon the worldwide stock of information, scientific talent and material, and for establishment building to deal with several of the research desires within the country. The agricultural research system in Asian country has reached a stage in its development wherever it may take a additional active role in joint research with foreign scientists as equal partners in addition as in coaching scientists from alternative countries.

Global Agricultural Research System

It is system that supports and funds to hold out research to develop agriculture and its connected aspects to cut back hunger and poorness within the world. The consolatory cluster on International Agricultural Research (CGIAR) could be a part of world Agricultural Research System.

CGIAR (Consultative Group on International Agricultural Research)

The consolatory cluster on International Agricultural Research (CGIAR) was established in 1971. The CGIAR is associate association of public and personal members that support and fund variety of International Agricultural Research Centres (IARCs) that do Research to cut back hunger and poorness throughout a stable network within the world.

CGIAR could be a world partnership that unites organizations engaged in Research for property development with the funders of this work. The funders embrace developing and industrial country governments, foundations, and international and regional organizations. The work they support is dole out by fifteen members of the syndicate of International Agricultural Research Centers, in shut collaboration with many partner organizations, as well as national and regional research institutes, civil society organizations, academia, and also the non-public sectors.

Objectives

- Food for People: produce and accelerate property will increase within the productivity and production of healthy food by and for the poor;
- Environment for People: Conserve, enhance and property use natural resources and diversity to boost the livelihoods of the poor in response to global climate change and alternative factors; and
- Policies for People: Promote policy and institutional modification which will stimulate agricultural growth and equity to learn the poor, particularly rural girls and alternative underprivileged teams.

International Agricultural Research Centers

There are 15 centers supported by the Consultative Group on International Agricultural Research (CGIAR)

International Agricultural Research System

1. CIAT (International Centre for Tropical Agriculture)
2. CIFOR (Centre for International Forestry Research)
3. CIMMYT (International Maize and Wheat Improvement Centre)
4. CIP (International Potato Research Center)
5. ICARDA (International Center for Agricultural Research in the Dry Areas
6. ICRISAT (International Crops Research Institute for Semi-Arid Tropics)\
7. IFPRI (International Food Policy Research Institute)
8. IITA (International Institute of Tropical Agriculture)
9. ILRI (International Livestock Research Institute)
10. BDI (Bio-diversity International)
11. IRRI (International Ricc Research Institute)
12. IWMI (International Water Management Institute)
13. WAC (World Agro-forestry Centre) CENTRE
14. WFC (World Fish Center)
15. AfricaRice (Africa Rice Center)

CGIAR Gene Banks

CGIAR scientists play major roles in collecting, characterizing and conserving plant genetic resources. Eleven Centers together maintain over 650,000 samples of crop, forage and agroforestry genetic resources in the public domain.

A number of agricultural research centres aside from CGIAR ar being operated with equal international reputes

1. CIAT - International Center for Tropical Agriculture Computer Network.

Centro International Diamond State Agricultural Tropical
Headquarters: A.A. 6713, Cali, Colombia
ciat.cgiar.org

CIAT is associate degree agricultural research establishment. They specialize in scientific solutions to hunger within the tropics. They believe that eco-efficient agricultural developing property ways of food production as the most effective thanks to eradicate hunger and improve livelihoods within the region.

CIAT is additionally regarding partnerships. By operating along with likeminded organizations they make sure that they need a sweeping and lasting impact. CIAT cooperates with several national and international establishments in developing and industrialized countries. Their relationships with them not solely operate on a matched basis however conjointly take the shape of networks and consortia.

CIAT may be a leading not-for-profit organization and has been serving to smallholders grow additional food and earn extra money for forty years. Established in 1970, jointly of the four original research centers within the consolatory cluster of International Agricultural Research (CGIAR), CIAT currently works in additional than fifty countries worldwide.

CIAT's headquarters, at the 500-hectare Agronatura Science Park close to city, in southwest South American nation, employs a number of the most effective scientists within the world. Their consultants in regional offices in continent and Asia change them to succeed in poor farmers fleetly and effectively.

2. CIFOR - Center for International Forestry Research

Headquarters: P.O. Box. 6596, JKPWB, Jakarta, Indonesia
www.cifor.cgiar.org

The Center for International Biology Research may be a noncommercial, world facility dedicated to advancing human upbeat, environmental conservation and equity. Center conducts Research that permits additional hip and equitable higher cognitive process regarding the employment and management of forests in less-developed countries.

Center's Research and Skilled Research policy facilitate manufacturers and practitioners form effective policy, improve the management of tropical forests and address the requirements and views of individuals UN agency rely upon forests for his or her livelihoods. The multidisciplinary approach of center considers the underlying drivers of deforestation and degradation which frequently lie outside the biology sector: forces like agriculture, infrastructure development, trade and investment policies and enforcement.

3. CIMMYT - International Maize and Wheat Improvement Center

(The abbreviation "CIMMYT": Centro International Diamond State Mejoramiento de Maíz y Trigo) Headquarters: Apdo. communication 6-641 06600, Mexico, D.F., Mexico
http://www.cimmyt.org

CIMMYT may be a non-profit Research and coaching center headquartered in United Mexican States

Mission: To sustainably increase the productivity of maize and wheat systems to make sure world food security and cut back poorness.

Vision and strategic goal: CIMMYT works with and brings along public research and extension organizations, personal firms, advanced research institutes, NGOs, and farmer associations in countries worldwide, operating pragmatically and apolitically to fight hunger and poorness the middle applies the most effective science to develop and freely share:

- High-yielding, stress tolerant maize and wheat varieties.
- Large, distinctive collections of maize and wheat genetic resources.
- Productivity-enhancing, resource-conserving farming practices.
- Training and knowledge with reference to the on top of.
- Through these activities and outputs, CIMMYT works to foster world and native food security, serving to farmers meet speedily rising demand from increasing populations and wealth, whereas addressing the rising challenges of world temperature change and resource degradation and scarcities.
- CIMMYT achieves the on top of with specialized employees and five hundred support staff from regarding forty countries, on a yearly budget of some USD fifty million. The middle is funded by international and regional development agencies, national governments, personal foundations, and also the personal sector. It's a member of the consolatory cluster on International Agricultural Research (CGIAR).
- The successes of the new crop varieties, beside improved management practices just like the use of plant nutrients, sparked the widespread adoption of improved varieties and farming techniques within the developing world which coined as development that came to be referred to as the "Green Revolution." The social and economic advantages of this movement were recognized worldwide once the Nobel Peace Prize was awarded to Norman Borlaug in 1970. The subsequent year, a little cadre of development organizations, national sponsors, and personal foundations organized the consolatory cluster on International Agricultural Research (CGIAR) to unfold the impact of research to additional crops. CIMMYT was one in all the primary international research centres to be supported through the CGIAR.

4. CIP - International Potato Center
(It is thought by its Spanish word form, (CIP means that Centro Internacional Diamond State LA PAPA) http://www.cipotato.org/

Headquarters: P.O. Box 1558, Lima, Peru

The International Potato Center (known by its Spanish word form, CIP) seeks to scale back poorness and attain food security on a sustained basis in developing countries through research and connected activities on potato, sweet potato, different root and tuber crops, and on the improved management of natural resources within the range of mountains and different mountain areas.

CIP headquarters is in La Molina, outside of capital of Peru, Peru's capital, in associate degree irrigated coastal depression. CIP conjointly has experimental stations in Huancayo within the high range of mountains and in San Ramón on the jap, rainforest-covered slopes, taking advantage of Peru's varied geographics and climate. The middle has another high range of mountains experiment station in national capital, Ecuador, and a worldwide network of regional offices and collaborators.

CIP has recruited a world team of scientists from twenty five countries, supported by across the country recruited workers. In its initial year of operation, CIP was funded by 5 donors. Today, the Center's budget is underwritten by over forty donors.

A number of agricultural research centres other than CGIAR are being operated with equal international reputes.

5. ICARDA- International Center for Agricultural Research In Dry Areas

Headquarters: P.O. Box. 5466, Aleppo, Syria
http://www.icarda.cgiar.org/Facelift.htm

In 1972, the CGIAR commissioned a team of consultants to review the potential threats of food shortages and loss of natural resources within the dry areas. The team counseled that associate internationally supported, research facility be established to serve developing countries with giant dry areas. The planned center would specialize in sub-tropical (temperate) zones. ICARDA's beginning charter was signed in 1975, with 3 world organization agencies (the Food and Agriculture Organization, the Global Organization Development Program and therefore the World Bank) as co-sponsors, and Canada's International Development Research Centre because the corporal punishment agency. Syrian and Arab Republic provided a 948-hectare website the subsequent year, and operations began in 1977.

ICARDA's headquarters, Tel Hadya, on the outskirts of the historic town of Alep, is a perfect website, for several reasons. Biophysical conditions – topography, soils, precipitation – area unit typical of the world's non-tropical dry areas, ICARDA primary mandate space. It additionally lies within the heart of the geographical region, wherever agriculture began 10,000 years back, and wherever several of the world's most significant crops originated or were initial domesticated. Plant genetic diversity within the region is nearly distinctive and this diversity permits scientists to uncover new genes that management very important traits appreciate drought tolerance, malady resistance or grain quality.

6. ICRISAT - International Crops Research Institute for the Semi-arid Tropics http://www.icrisat.org/

Headquarters: Pattancheru 502 324, province, India

The International Crops Research Institute for the Semi-Arid Tropics (ICRISAT) may be a non-profit, non-political organization that conducts agricultural research for development in Asia and Black Africa with a large array of partners throughout the globe. ICRISAT is headquartered in Hyderabad, province, India, with 2 regional hubs and 4 country offices in Black Africa. It belongs to the syndicate of Centers supported by the consolatory cluster on International Agricultural Research (CGIAR). Covering 6.5 million sq. kilometers of land in five countries, the semi-arid tropics have over two billion folks, and 644 million of that area unit the poorest of the poor. ICRISAT and its partners facilitate empower these poor folks to beat impoverishment, hunger and a degraded setting through higher agriculture.

The semi-arid regions (dry tropics) covering fifty five countries of sub-Saharan continent and Asia and haunted by over 800 million folks, area unit the poorest places on earth. These regions are area unit most at risk of global climate change with little or no precipitation, degraded soils and poor social infrastructure. The International Crops Research Institute for the Semi-Arid Tropics (ICRISAT) is that the solely research facility within the world serving the semi-arid tropics, at the side of its partners.

7. IFPRI-International Food Policy Research Institute

Headquarters: 2033 K Street, Washington, DC., USA
http://www.ifpri.org/

IFPRI is a world agricultural research facility based within the early seventies to boost the understanding of national agricultural and food policies to push the adoption of innovations in agricultural technology. In addition, IFPRI was meant to shed a lot of light-weight on the role of agricultural and rural development within the broader development pathway of a rustic. per its web site, IFPRI "seeks property solutions for ending hunger and impoverishment."

IFPRI is an element of a network of international research institutes funded partly by the consolatory cluster on International Agricultural Research (CGIAR) that successively is funded by governments, personal businesses and foundations and therefore the IBRD.

The International Food Policy Research Institute (IFPRI) seeks property solutions for ending hunger and impoverishment. IFPRI is one amongst fifteen centers supported by the consolatory cluster on International Agricultural Research (CGIAR), associate alliance of sixty four governments, personal foundations, and international and regional organizations

8. IITA - International Institute of Tropical Agriculture

Headquarters: Oyo Road, PMB 5320, Ibadan, Nigeria
http://www.iita.org/

International Institute of Tropical Agriculture (IITA) is Africa's leading research partner find solutions for hunger, deficiency disease, and economic condition. IITA's triumph Research for development (R4D) addresses the event wants of geographical area. IITA works with partners to boost crop quality and productivity scale back producer and shopper risks and generate wealth from agriculture. IITA is noncommercial organization based in 1967, ruled by a Board of Trustees, and supported by many countries. IITA works on cowpea, soybean, banana/plantain, yam, cassava, and maize.

Research-for-development platform: IITA's mission is to boost food security and improve livelihoods in Africa through research for development (R & D). IITA uses the R&D model in setting a groundwork course that addresses major development issues in Africa instead of merely causative to knowledge domain. It's tested to be an efficient mechanism for agricultural

ILRI works in partnerships and alliances with different organizations, national and international, in stock research, coaching and knowledge. ILRI works all told tropical developing regions of Africa and Asia.

9. Internatioanl Livestock Research Institute

ILRI could be a member of the pool of CGIAR centers that conducts food and environmental research to assist alleviate economic condition and increase food security whereas protective the resource base.

ILRI works on: ILRI's research is directed to handle seven international stock challenges. :

- Vaccine and diagnostic technologies for orphan animal diseases
- Animal genetic resources
- Climate change- adaptation and mitigation
- Emerging diseases
- SPS and market access among border market opportunities for the poor

- Sustainable intensification in sodbuster crop-livestock systems
- Vulnerability of marginal systems and folks

Capacity building at ILRI: ILRI's capability building initiatives ar supported 5 principles:

- Broad participation and a client-driven agenda;
- Building on native capabilities;
- On-going learning and adaptation;
- Long term commitment and partnerships; and,
- Integration of activities to handle complicated issues

10. Bioversity International

http://www.bioversityinternational.org/

Headquarters: Maccarese, Italy, simply outside Rome

Bioversity International is that the world's leading organization dedicated to agricultural multifariousness research to enhance people's lives through:

- Better nutrition, particularly in developing countries
- Sustainable farming practices, to secure our future food provides
- Conservation and use, to make sure that everybody will grow the food they have
- Research is that the core activity of IWMI. Research agenda is organized around four priority Themes as well as Water convenience and Access; Productive Water Use; Water Quality, Health and atmosphere and Water and Society. Cross cutting activities all told themes embody, assessment of land and water productivity and their relationship to economic condition, identification of interventions that improve productivity further as access to and property of natural resources, assessment of the impacts of interventions on productivity, livelihoods, health and environmental property.

IWMI works through cooperative research with several partners within the North and South and targets policy manufacturers, development agencies, individual farmers and personal sector organizations.

11. IRRI - International Rice Research Institute http://irri.org/

Headquarters: DAPO Box 7777, Metro Manila 1301 Philippines

The International Rice Research Institute (IRRI) is an autonomous, nonprofit, agricultural research and training organization. Its main goal is to find sustainable ways to improve the well-being of present and future generations of poor rice farmers and consumers while at the same time protecting the natural environment.

12. IWMI - International Water Management Institute

Headquarters: Colombo, Sri Lanka
http://www.iwmi.cgiar.org/

IWMI is one of 15 International research centers supported by the network of 60 governments, private foundations and international and regional organizations collectively known as the Consultative Group on International Agricultural Research (CGIAR). It is a non-profit organization with a staff of 350 and offices in over 10 countries across Asia and Africa and Headquarters in Colombo, Sri Lanka.

IWMI's Mission is to improve the management of land and water resources for food, livelihoods and the environment.

IWMI's Vision, reflected in the Strategic Plan is water for a food-secure world.

IWMI targets water and land management challenges faced by poor communities in the developing world/or in developing countries and through this contributes towards the achievement of the UN Millennium Development Goals of reducing poverty, hunger and maintaining a sustainable environment. These are also the goals of the CGIR.

Research Ethics: The pursuit of our mission requires carrying out research that involves people and the environment, and this has consequences both directly through the research activity and indirectly through the results. We therefore have ethical considerations to take into account when planning and implementing research. This policy sets out the principles and standards the World Agroforestry Centre has adopted to guide our work.

Intellectual property: The pursuit of our mission requires generating data and knowledge products of many types, including: books, papers and briefs, databases, maps, models, software and data archives. All of these are essentially international public goods. They also represent intellectual property (IP) that has to be managed and disseminated if they are to lead to changes in the lives of the rural poor. This document sets out the guiding principles we have adopted.

13. World Agroforestry Centre

Headquarters: United Nations Avenue, Gigiri, PO Box 30677, Nairobi, 00100, Kenya
http://www.worldagro for estrycentre.org/

The World Agroforestry Centre is an element of the alliance of the advisory cluster on International Agricultural RESEARCH (CGIAR) centres dedicated to generating and applying the most effective on the market data to stimulate agricultural growth, raise farmers' incomes, and shield the atmosphere.

The Centre's vision could be a rural transformation within the developing world as sodbuster households strategically increase their use of trees in agricultural landscapes to enhance their food security, nutrition, income, health, shelter, energy resources and environmental property.

The Centre's mission is to come up with science-based data concerning the varied roles that trees play in agricultural landscapes, and use its RESEARCH to advance policies and practices that profit the poor and therefore the atmosphere.

The World Agroforestry Centre is guided by the broad development challenges pursued by the CGIAR. These embody economic condition alleviation that entails increased food security and health, improved productivity with lower environmental and social prices, and resilience within the face of global climate change and different external shocks.

Headquartered in national capital, Kenya, the centre operates 5 regional offices placed in Bharat, Indonesia, Kenya, African nation and French Sudan and conduct RESEARCH in eighteen different countries round the developing world.

World Agroforestry Centre receives funding from over fifty totally different investors; as well as governments, personal foundations, international organizations and regional development banks. World Agroforestry Centre's work is conducted with partners from a variety of scientific and development.

Role of World Agroforestry Centre: The world population has now surpassed 6.5 billion. Demand for food increases as populations grow. Meanwhile, over one billion people continue to endure lives of extreme poverty. Agroforestry is uniquely suited to address the need to grow more food and biomass for fuel while sustainably managing agricultural landscapes for the critical ecosystem services they provide. It can serve as a means of curbing greenhouse gas emissions by slowing forest conversion to farmland and sequestering more carbon in trees on farms.

With over three decades of work with smallholder farmers in Africa, Asia and Latin America, and strategic alliances with advanced laboratories, national research institutions, universities and non-governmental organizations, the World Agforestry Centre is uniquely positioned to address global challenges.

To improve the livelihoods of poor smallholders and improve the sustainability and productivity of agricultural landscapes, we are:

- Broadening the range and diversity of trees that can be integrated into farming systems, especially as many produce higher income per unit of area than annual crops, require less labour and are more resilient to drought.
- Maximizing the productivity of agroforestry systems through improved tree germplasm, integrated soil fertility and the enhanced supply of high-quality tree fodder resources.
- Improving the income of poor households by facilitating their access to markets. This is also important in stabilizing land-use change in some areas, as well as increasing farmers' investment in agroforestry trees and systems.
- Working in agricultural landscapes that experience the greatest environmental stress to balance improved productivity with the sustainable management of natural resources. For example, stabilizing forest margins in Southeast Asia by converting slash-and-burn systems, and rehabilitating degraded agricultural land throughout Africa.
- Managing trees in agricultural landscapes to ensure the health of river and groundwater systems.
- Examining reward systems or other types of institutional and policy innovations (such as for carbon or water) to sustain biodiversity at the interface between smallholder agricultural landscapes and conservation areas.

14. World Fish Center

Headquarters: Penang, Malaysia
http://www.worldfishcenter.org/wfcms/HQ/Default.aspx

World fish center began in 1977 as the International Center for Living Aquatic Resources Management (ICLARM) based in the Philippines. In 2000 they shortened center's name to The World Fish Center and established new headquarters in Penang, Malaysia.

Mission: Mission is to reduce poverty and hunger by improving fisheries and aquaculture. We began in 1977 as the International Center for Living Aquatic Resources Management (ICLARM) based in the Philippines. In 2000 we shortened our name to The World Fish Center and established our new headquarters in Penang, Malaysia.

IRRI was established in 1959 with support from the Ford and Rockefeller Foundation and the Philippine Government. It is Asia's largest non-profit agricultural research center.

Most of IRRI's research is done in cooperation with the national agricultural research and development institutions, farming communities, and other organizations of the world's rice producing nations.

Its research activities began in 1962 and are now estimated to have touched the lives of almost half the world's population. The IRRI played a major role in sparking the Green Revolution on the field of rice when it developed its first new variety of Rice grassy stunt virus resistant rice in the 1960s. The Institute has a collection of 100,000 rice cultivars.

The Institute's research headquarters includes laboratories and training facilities on a 252 - hectare experimental farm on the lower campus of the University of Philippines in Los Banos Laguna, about 60 kilometers south of the Philippine capital, Manila. In addition to rice research, IRRI is also very active in local communities providing educational scholarships; organizing income-generating training activities and arranging other community projects that will help improve living conditions in the poor communities that neighbor the Institute.

IRRI was established to help poor rice farmers in developing countries grow more rice on less land using less water, less labor, and fewer chemical inputs. By helping to greatly boost production and ease the use of farm chemicals during its first 40 years, IRRI clearly showed the importance of rice and agricultural research in helping poor nations develop. The Institute's importance has been further reinforced by the private sector's traditional lack of interest in rice research.

15. Africa Rice Center (Africa Rice)
Headquarters: AfricaRice temporary headquarters is based in Cotonou, Benin
http://www.warda.cgiar.org/

The Africa Rice Center (AfricaRice) is a main container African research association with a mission to add to destitution lightening and sustenance security in Africa through research, improvement and organization exercises. It is one of the 15 global horticultural research Focuses upheld by the Consultative Gathering on Worldwide Farming Exploration (CGIAR). It is additionally a self-ruling intergovernmental inquire about relationship of African part nations.

The Middle was made in 1971 by 11 African nations. Today its participation contains 24 nations, covering West, Focal, East and North African locales, specifically Benin, Burkina Faso, Cameroon, Focal African Republic, Chad, Côte d'Ivoire, Law based Republic of Congo, Egypt, Gabon, the Gambia, Ghana, Guinea, Guinea Bissau, Liberia, Madagascar, Mali, Mauritania, Niger, Nigeria, Republic of Congo, Senegal, Sierra Leone, Togo and Uganda.

Perceiving the vital significance of rice for Africa and the powerful geographic extension of the Inside – which was constituted as the West Africa Rice Improvement Affiliation (WARDA) in 1971 – its Board of Clergymen took a notable choice in September 2009 to formally change its name to —Africa Rice Center (AfricaRice)‖ and to never again allude to it as WARDA.

Its main goal is to add to destitution mitigation and sustenance security in Africa, through research, advancement and association exercises went for expanding the efficiency and productivity of the rice division in ways that guarantee the maintainability of the cultivating condition. The business as usual of the Middle is association at all levels. Its innovative work exercises are led in a joint effort with different partners—basically the National Horticultural Exploration Frameworks (NARS), scholastic foundations, propelled examine establishments, ranchers' associations, non-administrative associations, and benefactors—for the advantage of African agriculturists, for the most part little scale makers, and also the a huge number of African families for whom rice implies nourishment.

The Inside hosts the African Rice Activity (ARI), the Local Rice Innovative Work System for West and Focal Africa (ROCARIZ), and the Inland Valley Consortium (IVC).

AfricaRice brief base camp is situated in Cotonou, Benin; inquire about staffs are likewise situated in Senegal, Nigeria, Tanzania and Côte d'Ivoire

Other International Centers

Earlier we discussed 15 international centers which are attached with Consultative Group on International Agricultural Research (CGIAR). Followings are some more international centers which are not directly attached with CGIAR but working in different fields.

ICLARM: International Center for Living Aquatic Resources Management

Headquarters: 11960, Bayan Lepas, P.O. Box. 500, GPO 10670, Penang, Malaysia www.iclarm.org

Focus: To improve the production and management of aquatic resources, for sustainable benefits to present and future generations of low-income producers and consumers in developing countries, through international multidisciplinary research in partnership with national agricultural research systems. The declining state and threatened sustainability of fisheries due to over fishing exacerbated with poverty and pollution, and the potential for increase in aquaculture production, call for research which includes understanding of the dynamics of coastal and coral reef resource systems and of integrated agriculture-aquaculture systems, investigating alternative management schemes in these systems, and improving the productivity of key species.

AVRDC: Asian Vegetable Research and Development Centre
Headquarters: P.O. Box. 42, Shanhua, Tainan 741, Taiwan
www.netra.avrdc.org.tw/docs/intro.html

Focus: To enhance the nutritional well being and raise the incomes of poor people in the rural and urban areas of developing countries through improved methods of vegetable production, marketing and distribution, which take into account the need to preserve the quality of the environment.

Training / Fellowships: Training in vegetable research and development, and IPM

CATIE: Centro Agronomico Tropical De Investigacion Y Ensenanza (Centre for Research and Learning in Tropical Agronomy)

Headquarters: P.O. Box. 7170, Turrialba, Costa Rica
www.catie.ac.cr

Focus: To promote and stimulate research and technical cooperation in animal, plant and forest production, with the objective of providing alternatives for satisfying the needs of the American tropics, especially the countries of the Central American isthmus and the Caribbean.

Training / Fellowships: Training is a part of the commitment of CATIE to the development of agriculture and renewable natural resources of the member countries and orientates its action towards the requirement of national and regional institutions. Collaborative M.Sc and Ph.D. programmes are offered with the U.S., U.K. and German Universities.

IBSRAM: International Board for Soil Research and Management
Headquarters: P.O. Box. 9-109, Bangkok, Thailand
www.ibsram.org

Focus: To assist and speed applications of soil science in the interest of increasing sustainable food production in developing countries.

ICIPE: International Centre of Insect Physiology and Ecology

Headquarters: Nairobi, Kenya
www.icipe.org

Focus: The mandate of ICIPE is to (I) research in integrated control methodologies for crop and livestock insect pests and for insect vectors of tropical diseases and (II) strengthen the technological capacities of the developing countries in insect science and its application through training and collaborative work.

IFDC - International Fertilizer Development Centre

Headquarters: P.O. Box. 2040, Muscle Shoals, Alabama, 35662 USA
www.ifdc.org

Focus: To help the developing countries solve their food-deficit problems by focusing on the development of fertilizers and fertilizer practices to meet the special needs of their tropical and subtropical climates and soils.

Training / Fellowships: Training on knowledge of Fertilizer industry in developing the skill of managers and professionals

CARDI: Caribbean Agricultural Researches and Development Institute

Headquarters: University Campus, St. Augustine, Trinidad, and West Indies
www.cardi.org

Focus: To accelerate sustainable agricultural development through strategic management of those processes that generate, transfer, adapt and commercialize appropriate technology, which will improve the social and economic well being of Caribbean peoples.

TSBFP: Tropical Soil Biology and Fertility Program

Headquarters: Nairobi, Kenya
https://www.cbd.int

Focus: To contribute to human welfare and the conservation of the environment in the tropics by developing improved practices for sustaining tropical soil fertility through the management of biological processes and organic resources, in combination with judicious use of inorganic inputs.

ICIMD: International Centre for Integrated Mountain Development

Headquarters: 4/80, Jawalakhel, G.P.O. Box. 3226, Kathmandu, Nepal
www.icimod.org

CABI: Cab International

Headquarters: Newsworthy Way, Wallingford, Oxon 8DE, UK
www.cabi.org

Focus: To collect, analyze and disseminate information on agriculture, forestry, the management of natural resources and related science including human nutrition and health.

ACIAR: Australian Centre for International Agricultural Research

Headquarters: Traeger Court, Fern hill Park, Bruce ACT 2617, Australia
www.aciar.gov.au

Focus: ACIAR directs it to mobilize Australia's research capacity to help solve agricultural research problems of developing countries. Thus the Centre allocates about three-quarters of its research and development budget to promoting bilateral development-related research collaboration between Australia and individual developing countries. The remaining quarter provides Australia's contribution to the institutes of the international agricultural research system.

Training / Fellowship: ACIAR awards 10 fellowships each year to developing country scientist involved in ACIAR Projects to undertake Master or Ph.D. training at Australian Universities.

CLIMA: The Cooperative Research Centre for Legumes In Mediterranean Agriculture

Headquarters: The University of Western Australia, Ned lands, Perth, WA 6907
www.general.uwa.edu.au/u/climaweb/

Focus: It focuses on legumes for sustainable agriculture in the Mediterranean climate of southern Australia. Research and training cover a wide range of disciplines, genetic engineering of legumes and bacteria to assessment of farmer attitudes about risk and a new species.

ILRI: International Institute for Land Reclamation and Improvement

Headquarters: Wageningen, Netherlands
www.ilri.nl

Focus: To undertake applied research on the sustainable development of irrigated agriculture. To hold annual, post-graduate training courses on irrigation, drainage and related subjects. To provide technical support and specialists advisory services to irrigation and drainage projects abroad.

THE WB: World Bank

Headquarters: 1818 h Street, NW, Washington DC.
www.worldbank.org

Focus: To fight poverty with passion and professionalism lasting results and to help people help themselves and their environment by providing resources, sharing knowledge, building capacity and forging partnership in the public and private sector.

CIDA: Canadian International Development Agency

Headquarters: 200 Promenade du Portage, Hull, Quebec, KIA 0G4, Canada
www.cida.gc.ca

Focus: It supports sustainable development activities in order to reduce poverty and to contribute to a more secure, equitable and prosperous world.

JIRCAS: Japan International Research Centre for Agricultural Sciences

Headquarters: Tsukuba, Japan
www.jircas.affrc.go.jp

Focus: To promote the advancement of agriculture, forestry, and fisheries in developing regions of the world through integrated collaborative research programmes. The new research strategy is to develop production and utilization systems in sustainable agriculture, forestry and fisheries in harmony with the environment by conducting research on such topics. To rehabilitate, maintain and improve the utilization of natural resources, with emphasis placed on tropical forest and coastal eco-systems.

CIRDAP: Centre On Integrated Rural Development for Asia and the Pacific

Headquarters: Dhaka, Bangladesh
www.cirdap.org.sg

Focus: To assist national action, to promote regional cooperation, to act as a servicing institution for its member countries for promotion of integrated rural development through research, action research, pilot project, training and information dissemination. Amelioration of rural poverty in the Asia-Pacific region has been the prime concern of CIRDAP. The programme priority of CIRDAP is set under 4 areas of concern; agrarian development, institutional / infrastructure development, resource development including human resources and employment.

ODI: Overseas Development Institute

Headquarters: 111, Westminster Bridge Road, London SE2 7 JD, UK.
hptt:/www.odi@odi.org.uk

Focus: Our mission is to inspire and inform policy and practice which lead to the reduction of poverty, the alleviation of suffering and the achievement of sustainable livelihoods in developing countries. We do this by locking together high-quality applied research, practical policy advice, and policy-focused dissemination and debate. We work with partners in the public and private sectors, in both developing and developed countries.

Training / Fellowships: The ODI Fellowship Scheme places up to twenty young economists a year on attachment to the governments of developing countries. There are currently 40 Fellows working in 17 countries in Africa, the Caribbean and the Pacific.

IIED: International Institute for Environment and Development

Headquarters: 3 Endsleigh Street, London WC1H ODD, UK.
www.iied.org

Focus: IIED aims to provide expertise and leadership in researching and achieving sustainable development at local, national, regional and global levels. In alliance with others we seek to help shape a future that ends global poverty and delivers and sustains efficient and equitable management of the world's natural resources

IDRC: International Development Research Centre

Headquarters: Canada
www.idrc.org

Focus: To help communities in the developing world finds solutions to social, economic and environmental problems through research.

FAO: Food And Agriculture Organization

Headquarters : Viale delle Tesme di Caracalla, 00100, Rome Italy
www.fao.org

Focus: To raise levels of nutrition and pursuit of food security and standard of living, to improve agricultural productivity and to better the condition of rural population.

Site for International fellowship for scientific mobility

Conclusion

International agricultural research setup is required to meet global demand for continuous need of location specific new agricultural technologies and techniques be developed to keep pace with the growing stipulation for food (crops, livestock, fish and forest products) either without price increases or deteriorating natural resources and to increase in total resources devoted to agricultural research globally. Such improvements would require major institutional reform, including correcting major flaws in shared agricultural research institutions in developing countries and meaningfully involving the clients of research, i.e. farmers and other natural resource users, individually and collectively, within the selection of research priorities, the conduct and continuing evaluation of research programmes and therefore the validation of research results with effective partnerships, domestically, with universities, private firms and other entities involved.

The objective is also to deal with issues and constraints which would be effectively addressed through regional collaborative efforts, for instance the management of natural resources that transcend national boundaries. Institutions and organizations concerned with the knowledge system are reformed, in order that information is widely shared, research priorities reflect the real needs of stakeholders, responsibilities for generating and disseminating technologies rest at the extent most appropriate to realize results, and existing resources are used more effectively. Support for the international agricultural research system must be strengthened to supply the support needed by developing-country under National Agricultural Research System (NARS). Investments in Consultative Group of International Agricultural Research (CGIAR) should increase to facilitate its mission of promoting sustainable and efficient agriculture for food security, poverty alleviation and environmental protection.

Some Pertinent Short Questions

1. What is full form of CGIAR and its role?
2. What is IRRI and its research mandate?
3. What is CIRDAP and where headquarter is located?
4. What is CIMMYT and where headquarter is located?
5. Write full form of ICARDA and ICRISAT with their research mandate;

References

http://beasiswascholarships.com/search/scholarship%20and%20fellowship%20for%20scientific%20mobility%20in%20india

http://beasiswascholarships.com/search/scholarship%20and%20fellowship%20for%20scientific%20mobility%20in%20india

19

References

Ashraf J. 1985. The antiquity of tobacco (*Nicotiana tabacum*) in India, *Indica***22:** 91-101.

Aswal, J.S., Kumar, J and Shah, B. 2010. Evaluation of biopesticides and plant products against rice stem borer and leaf folder. Journal of Eco-friendly Agriculture.5(1): 59-61.

Barapatre, A. and Lingappa, S. 2003. Larvicidal and antifeedant activity of indigenous plant protection practices for Helicoverpa armigera (Hub.). Proc. Nation. Symp. Fronterier Areas Ent. Res., November 5-7, 2003, I.A.R.I. , New Delhi.

Bhishagratna K K L (Ed). 1907. An English translation of the Sushruta Samhita, Vol. I-Sutrasthanam, Calcutta.

Birthal, P.S., Singh, H. and Kumar,S. 2011. Agriculture, economic growth and regional disparities in India. Journal of International Developmen , 23, 119-131.

Bishop Carl Whitting. 1936. Origin and Early diffusion of the traction plough. *Antiquity, **10:*** 261.

Blench R. 2003. The movement of cultivated plants between Africa and India in prehistory, In: Food, Fuel and Fields: Progress in African Archaeobotany (Eds: Neumann K, Butler A and Kahlheber S) pp 273-292, Africa Praehistorica 15 Heinrich-Barth Institute Cologne.

Boomathi, N., Sivasubramanian, P., and Raguraman, S. 2006. Biological activities of cow excreta with neem seed kernel extract against Helicoverpa armigera (Hubner). Annals of Plant Protection Sciences.14:11-16.

Carter, G.F. 1977. A hypothesis suggesting a single origin of agriculture. p. 89–133. In: C.A. Reed (ed.), The Origins of Agriculture. Mouton Publ., The Hague.Encyclopedia Britannica. 1970.

Cate, J. R. and Hinkle, M. K.. 1994. Integrated pest management: the path of a paradigm. National Audubon Society. Washington, D. C.

Chakrabarti D K. 2004. Indus Civilization Sites in India: New Discoveries. Mumbai, Marg Publications, ISBN 81-85026-63-7

Chakravarti H P. 1970. “History of irrigation in ancient India”, *Proceedings Indian History Congress*, 32nd Session, 1970, p.150.

Chand, R., Garg, S. and Pandey, L. 2009. Regional variations in agricultural productivity - A district level study. Report of the National Professor Project, NCAEPR, ICAR, New Delhi.

Chand,R. and Parappurathu, S. 2012. Temporal and spatial variations in agricultural growth and its determinants. Economic and Political Weekly, 47(26 & 27), 55-63.

China Faust, M. and Timon, B. 1995.Origin and dissemination of peach.Hort. Rev. 17:331 379. Janick, J. 2001. Asian crops in North America. HortTechnology 11:510–513.

Chopra R N, Chopra I C, Handa K L and Kapur L D. 1958.Indigenous Drugs of India.2nd edition, Academic Publishers (reprinted 1982) Calcutta and New Delhi.

Chowdhury K A 1965. Plant remains from *Deh Morasi Ghundai*, Afghanistan”, *Anthropological Papers, American Museum, Natural History,* **50** (2): 126-131.

Cuperus, G. W. & R. C. Berberet. 1994. Training specialists in sampling procedures. In, Handbook of sampling methods for arthropods in agricultural. (Eds. L. P. Pedigo and G. D. Buntin), CRC Press, Boca Raton, Fl.

Dhandapani, N., Shelkar, U.R. and Murugan, M. 2003. Bio-intensive pest management (BIPM) in major vegetable crops: an Indian perspective Food, Agriculture and Environment Vol. 1(2): 333-339.

Dhavalikar M K. 1974. Subsistence pattern of an early farming, *Puratattva,* **7:** 44-45.

Fairservis W A. 1971. The Roots of Ancient India. London, George Allen & Unwin, pp. 324.

Fuller D Q. 2000. Fifty years of archaeobotanical studies in India: Laying a solid foundation. In: Indian Archaeology in Retrospect; III: Archaeology and Interactive Disciplines (Eds: Settar S and Korisettar R) pp 317-390, Oxford and IBH New Delhi.

Ghosh E.1932. Studies on Rigvedic Deities astronomical and meteorological, *Journ. Asiatic Society of Bengal*, ***28***:56

Ghosh, P. K. and Sahoo, Bijoylaxmi. (2011). Indigenous Traditional Knowledge. Orissa Review January -2011 Members of Orissa Secretariat Data Entry Operators' Association greeting with Shri Bijoy Kumar Patnaik, Chief Secretary on New Year's Day at Secretariat on 1.1.2011.

Ghoshal U N. 1930. Agrarian System in Ancient India. University of Calcutta.p. 2.

Gode P K. 1945. History of *Cariaka* as food for horses, *Ann. of Bha.Ori.Res.Inst.,***26:**89- 105.

Gupta, Hrideshwer. 2013. Agricultural diversification in India, International Journal of Advanced Research in Management and Social Sciences, 2(5): 221-231.

Gupta, M. P. 2005.Efficacy of Neem in combination with cow urine against mustard aphid and its effect on coccinellid predators. Natural Product Radiance, 4(2):102-106.

Hand book of Agriculture, (2007) published by ICAR.

Harlan J R and Stemler A B L. 1976. The races of sorghum in Africa, In: Origins of African Plant Domestication (Eds: Harlan JR, de Wet JMJ and Stemler ABL) pp 465-478, Mouton The Hague.

Herman C P. 1997. Harappan Gujarat the archeology-chronology connection *Paltorient* **22:** 77-112.

Heyman E W. 1976. "India: Land-Tenure as an instrument of soil destruction", In: Soil and Civilization, John Murray Publication, London, pp. 183-189.

Hussain, Majid (1982): "Crop combination in India a Study". Concept Publishing Company, New Delhi.

Isman, Murray B. 2006. Botanical insecticides, deterrents, and repellents in modern agriculture and an increasingly regulated world. Annual Review of Entomology, Vol. 51: 45-66.

Jha B, Tripathy A. and Mohanty B. 2009. "Drivers of Agricultural Diversification in India", IEG Working Paper Series No.E/-/2009, Institute of Economic Growth, New Delhi.

Johannessen C L and Wang Siming. 1998. American crop plants in Asia before AD1500 *Pre-Columbiana*: *A Journal of Long-distance Contacts* **1:** 9-36.

Kajale M D. 1996. Palaeobotanical investigations at Balathal: preliminary results *Man Environ* **11:** 98-102.

Kanojia, A. K., Singh, R. V. Bambawale, O. M. and Trivedi, T. P. (Edr) 2005. In: Explored Indigenous Technical Knowledge in Pest Management: NCIPM, New Delhi.

Kareemulla, K, and Ravichandran, S. (2018). 105th FoCARS, ICAR-NAARM, Rajendranagar, Hyderabad, Telangana, India.

Karthikeyan, C. Veeraragavathatham, Karpagam, D. and Ayisha Firodouse, S. 2006. Cow based indigenous Tachnologies in dry farming. Indian Journal of Traditional Knowledge 5(1): 47-50.

Koohafkan Parviz and Altieri Miguel A. 2011. Globally important agricultural heritage systems: a legacy for the future. FAO, Rome, pp.41

Kosambi D D. 1975. An Introduction to the Study of Indian History, Popular Prakashan (3nd Revised Edition), Bombay, 1975, p.138.

Kulshrestha V P. 1985. History and ethnobotany of wheat in India *Journal d'Agriculture Traditionelle et de Botanique Appliquee* **32:** 61-71.

Lee, J-M.and Oda, M. 2003. Grafting of herbaceous vegetable and ornamental crops. Hort. Rev. 28:61– 124.

Li, J. and Yuan, L. 2000. Hybrid rice: Genetics, breeding, and seed production. Plant Breed. Rev. 17:15– 120.

Majumdar Ramesh Chandra and Pusalker A D (eds). 1951. The History and Culture of the Indian People. Volume I, The Vedic age. Bharatiya Vidya Bhavan, Bombay, India.

Max Muller, 1973.Biographies of words in: The Earliest Aryan Civilization, Collected Works, vol. X, pp. 134-135.

Md. Hedayetullah and Parveen Zaman. (2018). Forages of the World, Major Forage Crops. CRC press, Taylor and Francis Group, Inc., New Jersey, USA. ISBN: 13:978-1-77188-684-0.

Mehta, P K. 2009. Role of Crop Diversification in Output Growth in India: A State-Level Analysis, The IUP Journal of Agricultural Economics, 10(2): 24-42.

Mishra Renu and Sinha Ashutosh. 2014. Crop diversification in Indian Agriculture, Asian Journal of Research in Social Sciences and Humanities, 4(8): 113-120.

Mishra Renu and Sinha Ashutosh. 2014. Crop diversification in Indian Agriculture, Asian Journal of Research in Social Sciences and Humanities, 4(8):: 113-120.

Mittre V and Tiwari R K. 1978. Origin of Peasantry, *The Eastern Anthropologist,* **31**(3):219.

Mittre Vishnu and Savithri R. 1975-1976. Ancient Plant Economy at Inamgaon, *Puratattva,* **8**:55-62

Mudigourdra, S., Shekharappa and Balikai, R.A. 2009. Evaluation of plant products in combination with cow urine and panchagavya against sorghum shoot fly, Atherigona soccata Rondani Karnataka J. Agric. Sci., 22(3-Spl. Issue) : 618-620.

Mugisha Kamatenesil, M. Deng, A.L., Ogendo, J.O., Omolo, E.O., Mihale, M.J., Otim, M., Buyungo, J.P. and Bett, P.K. 2008. Indigenousknowledge of field insect pests and their management around lake Victoria basin in Uganda. African Journal of Environmental Science and Technology. 2(8): 342-348.

Nath Prana. 1931. Date of compilation of Kautilya's *Arthasastra", Indian Antiquary* ,**60**: 111.

Nawrot, J., Koul, O., Isman, M.B. and Harmatha, J. 1991. Naturally occurring antifeedants: effects on two polyphagous lepidopterans. J. Appl. Entomol. 111: 194-201.

Nayak, P. 1998. Unbalanced agricultural development in North-East India. Journal of Assam University, 3(1), 24-32.

Nene, Y.L. 2007. Scope of organic wastes,as eco-friendly materials in crop protection. Indian Journal of Plant Protection 35(2):161-167.

Oparacke, A.M., Dike, M.C. and Amatobi, C.I. 2006. Insecticidal Efficacy of SABRUKA Formulations as Protectants of Cowpea Against Field Pests. Journal of Entomology3(2):130-135.

Patanaik, H. P. 2017. Indigenous technical knowledge (ITK) in pest mananagement for sustainable agriculture. ENVIS Centre of Odisha's State Environment. Ministry of Environment, Forests & Climate Change, Govt of India.

Peries, I. 1989. Cattle urine as a substitute for Agrochemicals.National Rural Conference. In: Natural Crop Protection in the Tropics. AGRECOL, Publication, Okozentrum, Switzerland. pp 188.

Poonam, C. 2003. Effectiveness of cow urine and its decoctions against insect pests of soybean. M. Sc. Thesis, G.B.P.U.A.T., Pantnagar, pp. 95.

Pradhan, H. 2011. Validation of a few ITKs' for the control of major insect pests of brinjal. M.Sc (Ag) thesis submitted to OUAT, Bhubaneswar.

Prakash, A. and Rao, J. 1997: Botanical pesticides in agriculture. CRC Press. USA.

Prakash, N., Roy S. S. and Ngachan, S. V. (2013). Role of ITK in Conservation Agriculture: Blending Indigenous and Scientific Knowledge. ICAR Research Complex for NEH Region, Umiam – 793103, Meghalaya

Puhvel J. 1964. The Indo-European and Indo-Aryan Plough - a linguistic study of technological diffusion, *Science and Technology.* *5*(2):176-189.

Pullaiah T (2002) Medicinal Plants in India, 2 vols. Regency Publications New Delhi.

Raghuprasad, K.P., B.M. Akarsha and K. Raghavendra 2012."Raitha Samparka Kendras and Their Role in Agro-Information Delivery" Karnataka Journal of Agricultural Sciences, Vol.25, No.1, pp.82-85.

Raju, S.S., S. Parappurathu, R. Chand, P. Kumar and S. Msangi.2011. Biofuels in India: Policy, Potential and Emerging Paradigms, National Centre for Agricultural Economics and Policy Research, New Delhi.

Ramanujam S. 1976. Chickpea. In: Evolution of Crop Plants (Ed: Simmonds NW) pp 157-158, Longman Press, London.

Ramdas L A. 1950. Rainfall and agriculture, *Indian J. of Meteorology and Geophysics,* **4:** 268.

Randhawa. M.S.: "A History of Agriculture in India". Vol. 1, p. 23.

Rao, Shikaripura Ranganatha. 1991. Dawn and Devolution of the Indus Civilization, New Delhi, Aditya Prakashan, ISBN 81-85179-74-3.

Ratnagar, Shereen. 2006. Understanding Harappa: Civilization in the Greater Indus Valley. New Delhi: Tulika Books. ISBN 81-89487-02-7.

Raychaudhuri S P and Roy Mira. 1993. *Agriculture in Ancient India* – A Report, ICAR, Delhi, 1993, p.1;

Reddy, A.A. 2013. Agricultural productivity growth in Orissa, India: Crop diversification to pulses, oilseeds and other high value crops. African Journal of Agricultural Research, 8 (19), 2272-2284.

Sadwarte, A. K. and Sarode, S. V. 1997. Effect of neem seed extract, cow dung, cow urine alone and in combination against pod borer complex of pigeonpea. International Cheickpea and Pigeonpea Newsletter, 4: 36-37.

Saha Joydeep. 2013. Crop diversification in Indian agriculture with special reference of emerging crops. *Transactions*, 35(1): 139-146.

Saha Joydeep. 2013. Crop diversification in Indian agriculture with special reference of emerging crops. *Transactions*, 35(1): 139-146.

Sakhinetipalli, A. 2012. Field testing of ITK and insecticide based strategies for the control of brinjal fruit and shoot borer, Leucinodes arbonalis Guenee in brinjal.M.Sc(Ag) thesis submitted to OUAT, Bhubaneswar, during 2012.

Saraswat K S and Pokharia A K. 2003. Palaeoethnobotanical investigations at early Harappan Kunal *Pragdhara* **13:** 105- 139.

Saraswat K S. 1986. Ancient crop economy of Harappans from Rohira, Punjab (ca. 2000-1700 BC) *Palaeobotanist* **35:** 32-38.

Saraswat K S. 1992. Archaeobotanical remains in ancient culture and socio-economical dynamics of the Indian Subcontinent *Palaeobotanist* **40:** 514-545.

Sauer Jonathan D. 1993. Historical Geography of Crop Plants: A Select Roster. pp 15, CRC Press Boca Raton Florida.

Shailaja, B., Patnaik, H.P. and Mukherjee, S.K. 2012. Assessment of botanicals fermented in cow urine alone and along with panchagavya against brinjal shoot and fruit borer. Journal of Eco-Friendly Agriculture, 7 (1):24 – 28.

Sharma Y D. 1973. Value of Common Painted Ceramic Designs from Different Sites as Guide to Chronology with *Special* Reference *to* Pottery from Bara (Punjab), jn *Radiocarhon. and Indian Archaeology* (Ed. D.P. Agrawal and A. Ghosh), Tata Institute of Fundamental Research, Bombay.

Sharma, Vijay Paul and Thaker Hrima 2011. "Demand for Fertilizers in India: Determinants and Outlook for 2020", Indian Journal of Agricultural Economics, Vol.66, No.4, October-December, pp.638-661.

Shiyani R L and Pandya H R. 1998. Diversification of Agriculture in Gujarat: A Spatiotemporal Analysis, *Indian Journal of Agricultural Economics*, 53, (4): 627-639.

Shreth, C. N., Singh, K. I., William, S. J. 2009. Laboratory evaluation of certain cow urine extract of indigenous plants against mustard aphid, Lipaphis erysimi (Kaltenbach) infesting cabbage. Hexapoda. 16: 1, 11-13.

Shukla, A.N. and K. Kumar 1995. Restructuring Under-Graduate Curriculum in Extension Education for Manpower Development for Meeting Future Challenges of Agriculture, Paper presented at the National Workshop on Manpower Development in Extension Education for Meeting Future Challenges of Agriculture, National Academy of Agricultural Research Management, Hyderabad.

Singh Anurudh K (2017) Revisiting the status of cultivated plant species agrobiodiversity in India: An overview Proc Indian Natn Sci Acad 83 (1) 151-174 (DOI: 10.16943/ptinsa/2016/v82/48406).

Singh Jagadish: "India, A Comprehensive Systematic Geography". p. 32.

Singh, A.K. and Nigam, S.N. 2017. Ancient Alien Crop Introductions Integral to Indian Agriculture: An Overview, *Proceedings of the Indian National Science Academy,* 83 No. 3 September 2017 pp. 549-568.

Smartt J, Simmonds N W (Eds). 1995. Evolution of Crop Plants. p 531, Longman Group UK.

Somasekharan, J., Prasad, S. and Roy, V.P.N. 2011. Convergence hypothesis: Some dynamics and explanations of agricultural growth across Indian states. Agricultural Economics Research Review, 24(2), 211-216.

Sorenson J L and Johannessen C L. 2004.Scientific Evidence for Pre-Columbian Transoceanic Voyages, Sino-Platonic Papers, p 273, Department of East Asian Languages and Civilizations. University of Pennsylvania, Philadelphia, USA.

Srinivasan T M. 1970. Water-lifting devices in ancient India", *Indian Journal of History of Science*, **5**(2): 383.

Stansly, P. A., T. X. Liu, and D. J. Schuster. 2002. Effects of horticultural mineral oils on a polyphagous whitefly, its plant hosts and its natural enemies. pp. 120-133 In Spray Oils Beyond 2000 (edited by G. Beattie et al.).Univ. of Western Sydney Press.

State of Indian Farmers: A Report the Centre for the Study of Developing Societies (CSDS), Delhi conducted between December 2013 and January 2014.

Stern, V. M., R. F. Smith, R. Van den Bosch & K. S. Hagen. 1959. The integrated control concept. Hilgardia. 29:81-101.

Swain, M. 2002. Performance of agriculture in Orissa : Intertemporal and Spatial Variations in Orissa. Development Report, Planning Commission, New Delhi.

Swain, M., Swain, M. and Das, D.K. 2009.Regional disparity in agricultural development in Orissa in the pre and post reform period. Social Change, 39(2), 195-215.

Teaching Manual on Introductory Agriculture (Ancient heritage, Agricultural scenario Gender equity in Agriculture) *Compiled and edited by Prof. Janardan Singh,* Department of Agronomy, Forages and Grassland Management, College of Agriculture, CSK Himachal

Pradesh Krishi Vishvavidyalaya , Palampur 176062 India Published by: College of Agriculture CSK HPKV, Palampur 176062 H P June, 2015.

Thapar Romila. 2004. Early India: From the Origins to AD 1300. University of California Press, ISBN 978-0-520-24225-8.

Thomas P K. 1974. Zoological evidence from prehistoric India, *Bull.Deccan.Coli. Res. Institute* **34**:202.

Tripathy, U., Das, M. and Padhi, P.L. 2011. Regional variation in development profile of Orissa: An econometric analysis of indicators of development. International Journal of Multi-Disciplinary Research, 1(6), 107-130.

Ukey, S. P. and Sarode, S. V. 2001. Management of fruit borer and bud borer of chilli through integrated approach. Punjab Rao Krishi Vidhyapeeth Research Journal, 25 (1): 24-29.

Van der Werf, E. 1985. Pest Management in Ecological Agriculture. AME Foundation, Groenekan/Holland.Vijayalakshmi, K., Subhashini, B. and Shivani, V. K.,(1999). In: Plant in Pest Control - Garlic and Onion, Centre for Indian Knowledge System, Chennai, pp. 1-20.

Vavilov N I (1935) Theoretical Basis for Plant Breeding, Vol. 1. Moscow. Origin and Geography of Cultivated Plants. In: The Phytogeographical Basis for Plant Breeding (Transl: Like D) pp 316-366, Cambridge University Press Cambridge UK.

Verma, D.P.S. 2008). Secondary Agriculture: Value Addition to Primary Agriculture, submitted to Planning Commission, Government of India,

Vijayalakshmi, K., Subhashini, B. and Shivani, V. K.1999. In: Plant in Pest Control - Garlic and Onion, Centre for Indian Knowledge System, Chennai, pp. 1-20.

Vishnu-Mittre, Sharma Aruna and Chanchala. 1986. Ancient plant economy at Daimabad. (Appendix II) In: Daimabad 1976-79. Memoirs of the Archaeological Survey of India (Ed:Dali SA) pp 588-627, Government of India, Central Publication Branch, Calcutta.

Warren, D.M., L.J. Slikkerveer and D. Brokensha. 1995. The Cultural Dimension of Development: Indigenous Knowledge Systems. London: Intermediate Technology Publications.

Wright Rita P. 2009. The Ancient Indus: Urbanism, Economy, and Society, Cambridge University Press, ISBN 978-0-521-57219-4.

Zaman, A. and Maitra, S. (2019). Agricultural Heritage. Palmview publishing LLP, 189E, Kasba Road, Kolkata-700049, WB, India

Zaman, A. and Md. Hedayetullah (2019). Farming System and Sustainable Agriculture. *Agrotech Publishing Academy*. ISBN-978-81-8321-508-4.

Zaman, A., Zaman, P. and Maitra, S. 2017. Water resource development and management for agricultural sustainable knowledge. Journal of Applied and Advance Research, 2(2):73-77.